THE QUEST FOR SINGULARITY-
THE TWIN JOURNEY

THE QUEST FOR SINGULARITY- THE TWIN JOURNEY

SALIL KUMAR BHADURI

PARTRIDGE
A Penguin Company

ISBN:	Hardcover	978-1-4828-1310-4
	Softcover	978-1-4828-1311-1
	Ebook	978-1-4828-1309-8

To order additional copies of this book, contact
Partridge India
000 800 10062 62
www.partridgepublishing.com/india
orders.india@partridgepublishing.com

CONTENTS

The book is dedicated to my wife and son

PREFACE

In my childhood, the city life was not so much cramped and the sky-scrapers were not flexing their muscles so ruthlessly. There was abundance of blue sky and the eyes were not so much constrained to have a glimpse of that divine blue. I remember the thrill and enjoyment of staring at the night sky, lying on a bedcover stretched on the roof top of our building. The moving clouds and the twinkling stars in the vast and dark sky used to create such a sensation in the mind that even after so many years, whenever I see the boasting and power hungry humans with their so much limited capability and miniscule appearance compared to the mighty universe, it is only compassion that engulfs me and I go back to my memory of those days of feeling symphony of vastness and innocuous touch of the splendour in the tender mind.

My endeavour to write the book 'The quest for singularity-the twin journey' is in continuity to my thought developed in my childhood like every other child's inquisitiveness to know the surroundings.

I am grateful to those authors who are pioneers in their respective fields and from whose books I am amply benefited to accumulate information for writing this book. My indebtedness to them cannot be acknowledged by simply mentioning the names and their books. They are too great for any such acknowledgement. There are large number of references of articles published in internet in the acknowledgement pages through which I am enlightened to know so many things I was not aware.

Lastly, I must express my sincere thanks and gratitude to Dr. K.D Krori who has not only encouraged me in my effort but also has taken lot of pains to go through the chapters I, II, V and VI and helped me by correcting some of the contents.

S.K.Bhaduri
28.8.2013

CHAPTER I

Introduction

It was sometime before world war II when the extent of threats to mankind were still limited to war-cry of a few overenthusiastic and highly ambitious military powers. Yet the fear so created, caused an intense apprehension among other major powers of the world that lead them to think of adopting proper deterrent measures to safeguard their destiny from a complete debacle. Ultimately, the whistle of the world war game blew in September, 1939 and silently on a desperate move, United States took up their most scintillating and secret project—the Manhattan Project.

The silence broke in 1945 and it was the morning of the 16th July at 5:29:45 (Mountain War Time), at New Mexico's Alamogordo Bombing and Gunnery Range when the Trinity test was conducted with "The Gadget", the code name of the product of the Manhattan Project. The fireball shooting upwards with the characteristic mushroom cloud of radioactive vapour confirmed that the world had entered into Atomic Age, giving birth to the colossal blunder of the century and leaving several scientists involved in the project in acute despair for their failure in taking moral decision.

You may contemplate at this stage that I am inclined to describe the story of the first ever effort of a democratic country to achieve superiority over others by use of the most dangerous and mass killing weapons that massacred two modern cities of Japan at the end of world war II. No, not at all. Rather I have drawn the reference to alleviate the bandwagon of my thoughts on the subject I intend to describe in this book and am trying to search out answers to some of my basic queries.

Julius Robert Oppenheimer, an American theoretical physicist and Professor of Physics at Cal Tech, Berkeley, was often called the father of atomic bomb for his role in Manhattan Project. The brilliant light of explosion from the nuclear detonation test Trinity flashed the early morning skies with such intensity that Oppenheimer, being ecstatic, remembered the verse from the Bhagavad Gita and pronounced spontaneously, "If the brilliance of a thousand suns were to burst forth at once into the sky, that would display the magnificent strength of the Mighty One . . ." and "I am apostle of Death, the destroyer."

Bhagavad Gita, commonly referred as The Gita is an ancient Sanskrit text comprising of 700 verses of the Mahabharata (chapters 23-40 of Bhishma Parva) and is treated as a concise guide to Hindu philosophy. The content is the conversation between Lord Krishna and Arjuna in the battle of Kurukshetra in the beginning of the war. Krishna was responding to Arjuna's confusion and moral dilemma. While removing Arjuna's ignorance He explained that 'Time' and 'Death' would anyway engulf all those that Arjuna was worried for. By saying "Time am I" He clarified that He was Brahma or the Ultimate Truth. Then Lord Krishna blessed Arjuna with display of His divine absolute form and identity as "I am the Supreme, I am the time, the destroyer of worlds."

It has always been observed that exhibition of anything supernatural or superpower, not achievable by humans, is always attributed to some divine character/characters and who, as being considered as the creator of the universe, is believed to be responsible for controlling each and every conceivable activity in the universe by his desire. That is why Arjuna, the hero of the Mahabharata war, realized the absolute truth and could be convinced about the fate of the all living beings on earth by seeing the divine form of Lord Krishna. Identical situation happened to Oppenheimer. Even being highly educated in western culture and equipped with the extraordinary knowledge of modern science, he could not resist himself from expressing his inner spirit by chanting the same verses of the Gita by which the mythological hero, Arjuna was also inspired.

The quest for knowing the creator or the ultimate source of the creation had been a part of the journey of the human beings over the years. The queries of mankind got refined through intellectual thought processes and driven further to create two distinct ways of exploring the answers i.e. Philosophy and Science. When the curiosity could not find satisfactory answer through scientific experiments in the laboratory, they took shelter of imagination and went on searching for the answer. The philosophers have their conjectures and the scientists have their theories but no group has been able to grasp the ultimate reality through intellectual processes.

In Vedic philosophy, we find two prongs of knowledge. The knowledge obtained purely by observation i.e. the observable world which is the set of Achit. Another set of knowledge is entirely depending on ones state of consciousness and is called the set of Chit. The Achit set is a subset of Chit set because without Chit there will be nobody to observe. The balance part of the Chit set refers to the non-observable world giving rise to mysticism, mythology, intuition etc. Confronting this idea, the materialists say that Chit set is subset of Achit set and everything is explainable by physical laws and the consciousness grows purely from molecular interactions[24].

According to cosmology, the universe suddenly emerged from a singular source of infinite energy called space-time singularity-a state when all matter, energy, space, and time crushed into a point of infinite density. An identical view is expressed in Vedanta that the single source is the reality, consciousness and bliss. The singularity is a mathematical concept that brings together two mathematical concepts—Zero and infinity. Zero and infinity can be referred only qualitatively and they do not depend on other objects for their meaning i.e. they are absolutely independent. The identity of any number is lost when it is multiplied by zero or infinity. In Vedanta, zero and infinity are considered as unknowable and so explaining or defining them is beyond question. In mathematics also a variable can tend to approach them but cannot be arrived at. In fact it is not possible to find out any smallest number greater than zero i.e. if a number N is greater than zero, it cannot be the smallest number. Similarly if a number M is smaller than infinity, it cannot be the largest number. In Sanskrit, zero is more than a number, it is *Shunya* means void or emptiness. Also infinity is *Anadi* (without beginning), *Ananta* (without end). As per Vedanta, one has to transcend space-time to realize the ultimate reality, the true aspects of zero and infinity.

Zero and infinity come together by the singularity. Cosmic singularity means zero space-time and infinite energy. Vedanta says that during primordial state of existence of reality, there was infinite emptiness i.e. *Shunyata*. Unless one's mind is completely empty i.e. integrated to divine energy and consciousness and transcends space-time in deep meditation, it is not possible to realise this state. We and everything else in the universe are simply events in space-time. Anything having name and form has a beginning and end in space-time. Now the question is how can one transcend space-time? When we say 'one' it means not a physical body. Besides physical body, there is a subtle body. It is the subtle body that goes beyond space-time and experience the reality. When one thinks of an object, he separates himself from the object. So long as he, the subject and the object have separation and existence in space-time as two separate events he cannot transcend space-time. As subject-object duality vanishes, he gets merged with the universe. In this state, as per Upanisad, he sees everything in himself and himself in everything.

The separation between the observer and the observed or the duality is an illusion or Maya as identified by the Eastern mystics since long. Recently the scientific community also coming into the identical conclusion while explaining quantum mechanics.

While in micro-scale, the scientists are continuously involved in search for the fundamental particle, the most fundamental constituent of matter, in macro-world, the other section of scientific community is active in the discovery of the source of creation of the universe and the beginning of time. The intellectual pendulum which is swinging back and forth is to find the answer of the query— when did beginning of time take place? Was it simultaneous to the big bang? Or was the universe existing prior to that event? Thousands of questions rocked the minds of scientists and theologians over thousands of years, "Where do we originate from? What is our identity ? Which we are struggling for and so on ?" A story says that one day, the German philosopher Schopenhauer, being driven by his inner instinct of thoughts, was walking inattentively and entered unknowingly into a garden of somebody. Immediately, the guard rushed in and tried to stop him by saying, "Who are you?" Schopenhauer continued to walk as if he did not listen. The guard again asked him, "Where from you have come?" Still there was no response from the unmindful wise man. The guard then furiously interrupted him and said, "What for you have come?" Now it brought the philosopher to sense. Schopenhauer, most eagerly approached the guard and politely asked, "Hi man, through out my life, I am searching answers of these questions. I could not get the answers so far. Do you know the answers?"

In the cycle of birth, life and death, an effort of the individual was to discover his origin, identity and destiny and this also connected the thought process directly to cosmic ones. While doing so, the quest for singularity started tracing our ancestry linkages back through the generations, animal ancestors, primordial form of life, the basic ingredients of life and matter generated at the so-called beginning of the universe and vast ocean of energy pervading the space. Whether the singularity could be achieved extending such adventurous thinking forever backward? Whether all the roots are terminated at some point of time?

The pool of queries borne by the thinkers of the centuries saw the unveiling light in attempt of disclosure of the mysteries with Hubble's discovery in 1929 and the question of beginning of universe was brought into the realm of science and it was estimated that the singularity existed at a time about ten or twenty thousand millions years ago. The mathematical presentation of Penrose and Hawking proved in 1970 that there must have been a big bang singularity[1]. Again with developments in theoretical physics, especially after the inception of string theory the perspective has been changed. The pre-bang universe has now become the latest frontier of cosmology. In micro-level study, the elementary particle which is supposed to be the basic building block of the universe and from which all other particles are formed is the singularity. In recent experiment in CERN with huge accelerators, the scientists are in search for Higgs Boson, sometimes also called in light mood 'the God Particle'. It is believed to be the elementary particle which gave masses to other particles a fraction of a minute after the big bang, the initial explosion of the universe. In physical world, the singularity is referred to a quantity which is infinite. Also when the quantity approaches infinity, another parameter goes to zero. But zero is never achieved, for then the laws of physics will break down and therefore there must be a jump to zero and with apparent singularity going away, a new set of laws has to prevail. So If Higgs particle is an elementary particle, it cannot be found having been carrying no properties. The subject will be discussed in detail in the later chapters.

In this book I would like to describe the untiring efforts of the scientists and the philosophers for thousands of years to explore the truth behind the origin of the universe and the elementary particles of matter that constituted the whole of universe. One of the aspects is to understand the external reality through science in which process it tries to measure and describe reality after passing it through exhaustive scrutiny without personal, religious or cultural bias. The other one is the inner, personal experience of consciousness and awareness of awareness. This indistinguishable consciousness was described by the eastern spiritualists for thousands of years as the ultimate state of bliss or Nirvana. The concept of God in various religions did also build the foundations of their beliefs and attempted to explore the meaning of life as well as its origin.

Science constantly provides theoretical ideas and put the same for experimental verification and unless pass through successfully, the ideas are discarded however beautiful those may be. But whatever may be the outcome, the human society has always been benefited by the ideas in the way to rectify the flaws and come out with the correct one advancing over the debris of the discarded theories.

On the contrary, no religion has ever been rendered obsolete even though the facts and observations do not support its views. No religion embarks on scientific statements and even though scientists ridicule the claims of the religions, a believer of his religion has always observed the faith and does not get perturbed by any result of verification.

Einstein used to express his predilection by referring to God while talking about laws that relate the fundamental harmonies of the universe. But this God was definitely not the well imagined mystic personality of the religious minded people, but was the astounding, integrating super-activity running to unite the whole universe and that instigate the spiritual pyre in the minds of the human being. Paul Dirac, one of the greatest English theoreticians used to say about the beauty of mathematical equations that expressed the fundamental laws of the universe that this was almost a religion to him. Even at one time metaphorically he said in theological spirit, "God is a mathematician of a very high order."

It has been my effort to elaborate the twin aspects of the scientific exploration of the truth behind macro-world (the subject in chapter II related to the creation of the universe) and the micro-world (Chapter V relating to investigation and discoveries of fundamental particles of matter). At the same time, I have tried to elaborate the twin journey of the humanity to unveil the same truth, the absolute reality, by going beyond empirical approach following the process of self-realization (Chapter III) on the one hand and through intense beliefs on godhead (Chapter IV) on the other.

In order to conclude the introduction I would prefer to imagine that the two streams of thoughts either by scientific way of probing or expanding human spiritual senses through theological way of thinking, will continue their journeys for search of singularity and let their achievements supplement each other's efforts to place the civilization on strong footings ahead.

CHAPTER II

CREATION OF THE UNIVERSE– THE CONCEPTS

The scientific thought processes

Let us imagine the days when earth appeared to be flat to its first explorers, predators, hoarders, and inquisitive members of early civilisations. Obviously, what was beyond the boundaries of the flat earth was largely unknown and open to speculations. The starry heavens twinkling at a distance too far away in a clear sky were a source of endless wonder and inspiration to mankind to expand their imaginative power to find out the unexposed truth behind the origin of universe. Peoples from all parts of the world being inspired by the mighty presence of the endless skies and the innumerable celestial bodies gave birth to numerous myths. Their thirst for unveiling the mystery can be seen as an attempt to explain their own place in the universe. About six thousand years ago, the Sumerians believed that the Earth is at the centre of the universe. This belief was later percolated to the Babylonians and Greek civilisations.

From the very dawn of human civilization, humans tried to know the nature and the universe where it is embedded. Scientists as well as the philosophers probed into the secrets of nature and while doing so, the origin of the universe became the focal point of their investigations.

In the early days the thought of supernatural powers was so much imbibed in the minds of ancient thinkers like the Greek philosophers, that they were thinking that the universe was created by the super natural power of so called gods, but the they did not think it necessary to investigate how the gods accomplished the act of creation, for such matters were contemplated as divine and thus left outside man's intellectual reach. An excellent scientific picture of the universe was drawn by the Greeks who described the motions of the planets with precise mathematical formulations. They thought that those planets are like the stars and other celestial bodies which did never show any decay over the years.

In ancient China, the universe including everything on earth and in the sky, was imagined as part of a gigantic system. However, they could visualize a universe of many millions of years old and in this respect, they were close to today's view of age of the universe.

Western civilisation grew up under the influence of Greek ideas and also that of Chinese teachings, which insisted on a single God who was thought to be creator of the universe and sustaining it. This also resembled the idea as per Muslim faith. The Bible did not contain any scientific information about the universe and Galileo used to be fond of saying that the Bible taught the way how to go to the heaven and not the way heavens were existing. The Church did not allow any speculation on the matters related to established divine dictates to keep the masses under its control. It was a well known fact that the statement of Copernicus about the sun, and not the earth, moving around the spherical universe brought misfortune to him for the obvious reasons.

In the second millennium BC, the Sanskrit speaking Aryans migrated from central Asia to south Asia. They then infiltrated to northern part of Indian subcontinent through modern-day Afghanistan some time around 1500 BC. Within next 300 years, Rig Veda, the oldest of the Vedas was written in Vedic Sanskrit. By 1000 BC the Aryans began to settle down and coming in contact with the non-Aryan people formed the social bondage and Indo-Aryan culture. There were changes in thinking, diversity of opinions. The people became less interested in ritual routines and ritual sacrifice. The focus changed to the questions about self and its relationship to the universe. The new interest grew in introspective writings called the Upanisads or Vedanta, a collection of as many as two hundred books written over two centuries. The Upanisads became the foundation of Hinduism.

What was happening in Europe ? What ideas were being cropping up in that part of the world? Let us look into the concepts about celestial bodies which were being formalized and it was Aristotle who gave the idea about shape of earth in 340 BC and propounded that Earth was stationary and other planetary bodies including Sun were revolving around Earth. Thereafter, a complete cosmological model was proposed by Ptolemy in first century AD. Ptolemy's model was generally but not universally accepted but was adopted by Christian church as the true model of the universe since it matched with their scriptures[1].

In 1514, Copernicus presented his model with the Sun stationary at centre and Earth and other planets moving around the Sun. Kepler and Galileo, the two astronomers supported the model of Copernicus. Further Kepler modified the model with proposition of movement of planets around the Sun in elliptical orbits in place of circular orbit[1].

The most far reaching and revolutionary things happened in the field of science in 1664-1666 when Newton brought about the law of universal gravitation according to which each and every body in the universe was attracted towards each other with a force depending on the masses of the bodies and the distances between them.

At the early stage of cosmology, it was thought that universe started at a time not too far. In fact a date was proposed by St. Augustine which was about 5000 BC.[1] Whereas Aristotle and other Greek philosophers thought that universe has not been created. It was there and would exist forever. The periodic natural calamities set repeatedly the human race right back to the beginning of civilization.

With progress of science and philosophy two sets of contradicting ideas, beliefs and theories about the origin of the universe settled down. The Universe either existed eternally with no beginning or end or it was created at some point in time and would ultimately come to an end. I shall try to throw some light on the conception of creation of the universe from the scientific point of view and then on the context of early cultural, religious, and philosophical views of the beginning of the universe.

So far the questions of static or unchanging universe were concerned or whether there was a beginning of the universe was the subject of metaphysics or theology. The situation changed when Edwin Hubble as mentioned earlier, observed in 1929 that the universe is expanding and predicted that about ten or twenty thousand million years ago all the celestial bodies were at same place[1]. The discovery of the expanding universe was one of the revolutionary ideas of the twentieth century[1]. Hubble showed that the Milky Way galaxy which was a conglomeration of our Sun and innumerable stars are not the only constituents of the universe, there were large numbers of other galaxies and huge empty space between them. It is now known that a galaxy contains about hundred thousand million stars and there are about hundred thousand million galaxies out of which our galaxy is one of them. Can you now guess the size of our earth vis-à-vis the size of the universe! What a gigantic vastness is spread around us !

Steady State Theory

(No beginning No end)

In 1948 Fred Hoyle, Thomas Gould and Herman Bondi proposed a theory called Steady State Theory which stated that there was no beginning of the universe and there would be no end[1]. This theory is also called as Infinite Universe Theory or Continuous Creation Theory. As per the theory, the universe remained unchanged with time and obviously required that the new stars and galaxies would form at the same rate by which others were dying out to keep the density same all over the universe irrespective of time. The mix of old and new systems in the universe would remain same as we see it now. This proposition was necessary to counter the fact that the universe was expanding and therefore it was to be established that the decrease in density due to expansion of universe was compensated by the creation of matter in the form of new galaxies.

The problem in establishing the central idea of the theory showing the universe as static was tackled this way. The theory continued to be quite popular throughout the 1950s regardless the inherent problems associated with it. Even when Einstein formulated his general theory of relativity in 1915, he considered that the universe is static and modified his theory introducing a cosmological constant in the equations. His cosmological constant allowed space-time to expand to exactly balance the attraction of all the bodies in the universe to make the universe static.

The Steady State Theory began to fade in the 1960s after quasars were discovered. The Quasar is the quasi-stellar radio source and highly luminous galactic nucleus. It is a very compact zone in the centre of a super-massive black hole and highly energetic emitting very high energy. Quasars are at so much distance from the earth that their light take several billion years to arrive at the Earth! Hence quasars are objects from the past indicating the structure of the universe a few billion years ago which was obviously very different from what it is today.

Quasars discovered in 1963 found emitting extraordinarily powerful radio waves that hit the ideas of steady state theory. It was revealed that they were not just emitting radio waves, but also electromagnetic radiation which appeared to be coming from billions of light years away. The Steady State Theory indicated that both the radio waves and electromagnetic radiation from sources both far away and relatively closer should be available scattered across the universe. But it could not adequately explain this phenomena why quasars were located only in the most distant regions of the known universe.

The final blow to the Steady State Theory was delivered by two American radio astronomers Penzias and Wilson. These scientists were working for the design of a very sensitive microwave detector and accidentally discovered the cosmic microwave (high frequency radio wave) background which they thought to be left-over radiation from the Big Bang. The Steady State Theory explained this background radiation as the light from ancient stars which had been absorbed and emitted in all directions by particles from galaxies.

The detected microwave background was very smooth and most of the astronomers were of the opinion that it did not come from different small sources. Moreover the spectrum resembled that of an ideal black body which was supposed to absorb all electromagnetic radiation. Penzias and Wilson got Nobel prize in 1978 for their work[1]. The Steady State Theory is now no longer accepted by most cosmologists.

Oscillating Universe Theory

In subsequent discussions on expanding universe theories, I shall mention models based on the Russian mathematician Friedmann's assumptions and in one of the models there is indication that the expansion of the universe might have been followed by contraction. This is a oscillation between big bang and big crunch. That is, if the Big Bang describes the possible beginning of the universe, the Big Crunch explains how the universe will end as a consequence of that beginning. In that conception, our universe which we see now can be the first of a possible series of universes or it can be the nth universe in the series. We can think of both the Big Bang and the Big Crunch to simply make up a pair that mark the beginning and end, respectively, of one complete cycle of lifetime of the Universe called Big Bounce. In the oscillating universe, time is endless without any defined beginning.

The most recent measurements of the CMBR (cosmic microwave background radiation) by a very accurate measuring device known as the WMAP (Wilkinson Microwave Anisotropy Probe) shows that the Universe will continue to expand and will most likely end in what is known as a Big Freeze or Heat Death. The same device measured the age of our universe with sharp precision. It is therefore highly unlikely that future findings will deviate largely from what has been discovered about expansion of the universe. This expansion of the universe where the galaxies are drifting farther apart and accelerating the expansion of the universe is supposed to be due to one mysterious energy called as dark energy, which is considered to be responsible for this phenomena. We may have to shelve the

Oscillating Universe Theory, unless the actual properties of dark energy are very dissimilar from what it is showing now or there is some other form of matter, called as dark matter, not yet detected, which may raise the average density of the universe up to the critical value so that the expansion halts.

Dark Matter and Dark Energy

There is a mismatch observed between the gravitational mass and the luminous mass of galaxies and clusters of galaxies. While the gravitational mass of an object can be determined by applying laws of motion, the luminous mass is determined by adding up all the light and converting that amount to a mass applying the mechanism by which the stars glow. This mass-to-light ratio indicates that the energy in luminous matter contributes less than 1 per cent of the average energy density of the universe[18]

Therefore, there is certainly more matter in our galaxy and other galaxies that we cannot see or detect. We do not know 96% of the universe. The portion of the universe which is observable, i.e. the stars, galaxies, planets etc. comprises only 4 % of the universe. There is very strong arguments that at most 5 per cent of the mass and energy of the universe is in the form of atoms[18]. Thereby, the speculation about dark matter originates. The scientists think that during the evolution of the universe, the dark matter was responsible for material to clump together through gravity to produce the galaxies and the galaxy clusters. It appears very interesting that we are living in a universe in which we do not know 95% of its constituents. This is the most interesting and unresolved problem in astronomy and is known as the mystery of dark matter[19]. This is called dark because it cannot be seen or cannot be measured by any physical means. Scientists are of the opinion that about 23% of the universe is comprising of dark matter. This invisible mass holds the moving galaxies from flying apart.

The physicists and astronomers consider dark matter as some new particles which cannot be detected during particle accelerator experiments or in exploration of cosmic rays. It is anticipated that a new particle to behave as dark matter, must be heavy (perhaps heavier than a neutron) and weakly interacting with normal matter so that it does not easily lead to light-producing reactions. It is also considered that dark matter must be the basic building block of the largest structures in the universe i.e. various cosmic structures.

The concept of dark energy which is the expansive force, on the other hand has come up to explain the observed accelerated expansion of the universe. The theory of expanding universe cannot explain the acceleration without consideration of the effect of some additional unknown, undetected energy, which is called dark energy. It is called "dark" because, like dark matter, it may be very weakly interacting with regular matter and it is called as energy because it can be established that it contributes nearly 73 per cent of the total energy of the universe.

Fundamental Forces of Nature

Two extraordinary theories of the twentieth century that influenced the pace of progress of modern physics were the Quantum Mechanics and Einstein's theory of general relativity. While quantum mechanics opened the entry to the micro-world of sub atomic science i.e. secrets of nuclear physics, the other paved the way to the physics of the macro-world of the cosmic giants, the mystery behind creation of the universe, formation of black holes, dying stars etc. (see appendix B). Though these two theories, principally provided total knowledge of the universe, lack of compatibility between them puzzled the scientists. The efforts to unify these two theories were not successful, even Einstein was busy for last thirty years of his life to find out some bridging link between light and gravity by stipulation of some unified theory, combining the two basic forces of nature i.e. electromagnetic force and gravitational force.

The gravitational force, as we know, is the force that binds together the entire solar system including earth, the stars and the galaxies of the universe and everything is put in motion maintaining their tracks intact.

The electromagnetic force holds negatively charged electrons of the atoms with its positively charged nucleus. But inside nucleus, electromagnetic force is dominated by the strong force. The strong force binds together the protons and neutrons of the nucleus. The repulsive force between the positively charged protons is overpowered by the strong force dominant in the nucleus, otherwise the nucleus would have been ripped off. The balance between electromagnetic force and strong force in the nucleus has provided only about 100 odd elements since to maintain the above perfect balance between the two forces, the number of protons in the nucleus cannot be indefinitely large for then the repulsive force would have dominated over the strong force. When the strong force is unleashed from the nucleus of the atom, enormous energies are released. The sun and the stars which are similar to nuclear furnaces are glowing due to nuclear detonation[3].

The weak force is responsible for disintegration of the nucleus of elements having heavy nuclei into elements containing smaller number of protons[3]. This is called radioactivity. The release of energy due to weak force can produce enormous heat as it happens in the interior of earth where the disintegration of radioactive elements inside the earth's core generates heat.

Expanding phase of Universe and famous Big Bang Theory

Dated back to 1916, Albert Einstein noted that his field equations of general relativity predicted an expanding universe and that was the first direct scientific evidence for a big bang universe. But acceptance of such expansion of the universe meant that there was cosmic beginning. Einstein was unwilling to accept this. So he modified his theory to adapt common perception of his time, namely an eternally existing universe in conformation with the steady state theory.

Friedmann, the Russian mathematician took the general relativity at face value and tried to solve the problem with some basic assumptions that the universe is identical in all directions and it would be true even if it was observed from anywhere else[1].

There can be three different types of models on the basis of Friedmann's assumptions. One is that the expansion of universe is sufficiently slow and gravitational attraction between different galaxies can cause the expansion to slow down and ultimately stop. Once the expansion is stopped, the attraction between galaxies created movement towards each other resulting in contraction of the universe. The distance between them becomes zero and again increases to a maximum to get reduced again to zero[1].

Second is that the expansion is so fast that the gravitational attraction cannot stop expansion rather can slow down the speed to some extent with the net result of the separation between the galaxies starting from zero to more and more with time at steady speed[1].

Third model will be an expanding universe with separation between galaxies starting from zero and then increasing forever with slower and slower speed but never attaining zero value[1].

All the three models predict that at some point in the past, estimated as ten to twenty thousand millions years ago, the distance between adjacent galaxies must have been zero when density of the universe and space-time curvature would have been infinite. The general theory of relativity on which the Friedmann's models are based thus indicates that there was a singular point in the universe[1].

Light which was known to follow the features of electromagnetism propounded by Maxwell, was found not obeying the theories of mechanics of Newton when treated as particle. This phenomenon posed serious problem to the scientists of the nineteenth century. Michelson demonstrated in 1880s that it always travelled with the same velocity, regardless of the speed of its source. The problem was handled in various ways by the physicists. In 1892, FitzGerald an Lorentz independently attempted that by postulating that if the detector apparatus changed its size and shape in a characteristic way depending on its state of motion, the theory and experiment could be reconciled. Poincare, in 1898, suggested that intervals of time and the length, might be observer-dependent. In 1904, he further added that the speed of light might be an "unsurpassable limit"[17]. It was Einstein who at the age of sixteen viewed the problem in an unconventional way and wanted to imagine the situation if somebody travelled at the speed of light. He assumed that the laws of physics and the speed of light must be the same for all observers having uniform motion irrespective of their states of relative motion. To maintain this, space and time can no longer be independent. Rather, they are linked with each other in such a way as to keep the speed of light constant for all observers. Space and time are relative as they depend on the motion of the observer who measures them and light is fundamental independent of either. This is the basic idea behind Einstein's theory of special relativity. Minkowski announced in 1908 that "Henceforth space by itself, and time by itself, are doomed to fade away into mere shadows, and only a kind of union of the two will preserve an independent reality". Einstein did not initially agree to the four-dimensional interpretation of his theory by Minkowski and termed it as "superfluous learnedness" But he changed his mind quickly. The language of space-time proved to be essential in deriving his theory of general relativity[17].

Einstein observed the principle of equivalence (see Appendix A) and realized that a person accelerated downward along with a ball would not be able to detect the effects of gravity on it [17]. So when the person observed the ball, he could "transform away" gravity (at least in the immediate neighbourhood) simply by moving to this accelerated frame of reference — independent of type of object being dropped. So gravitation appeared (locally) equivalent to acceleration. If gravity behaved like other type of force, suppose electricity, the ball with more charge would have been attracted more to the earth and hence there would be no way to transform away such effects by moving to the same accelerated frame of reference for all objects. But gravity is "matter-blind" — it affects all objects the same way. From this fact Einstein drew the spectacular inference that gravity does not depend on the properties of matter. The phenomenon of gravity must be arrived at some property of space-time[17].

In Einstein's universe, the space-time is no longer flat in contrast to Newton's universe but warped by matter. Einstein ultimately identified that the curvature of space-time embodies the gravitational force. Mathematically, the curvature of space-time (see Appendix A) reflects the distance relations between its points. In most curved space-time, effect of gravity is strong and it has no effect where space-time is flat. This is the crux of Einstein's theory of general relativity, which can be concisely described as : Matter decides space-time how to curve, and curved space-time decides matter how to move.[17].

In the expanding universe, actually the galaxies are moving apart from one another with small sideways velocities and therefore they can only be very close together but not all at same place. So there are theories that say, in reality, the current expanding universe may not be formed from a big bang singularity, rather from an universe in contracting phase[1].

Whereas, three observations helped in concluding that the universe started with the big bang. First, the separations between galaxies were becoming larger and larger indicating that the universe is expanding. It was also concluded that everything in the universe used to be extremely close together before some kind of explosion. Second, the abundance of helium and other nuclei like deuterium (an isotope of hydrogen) in the universe perfectly explained big bang. A hot, dense, and expanding environment at the beginning could produce these nuclei in abundance as we notice today. Third, the cosmic background radiation from every direction in the universe assumed to be the afterglow of the explosion is actually observed by the astronomers[11].

It is possible to know whether a galaxy is approaching or receding by analysing the spectrum of light received from it. The spectral shift toward shorter wavelength (blue shift), indicates that the galaxy must be approaching towards us. If the shift is towards longer wavelength (red shift), then the galaxy must be receding. The degree of the shift depends on the speed of approach or recession. In other words, it is observed more galaxies whose spectrum was red-shifted than those whose spectrum was blue-shifted which indicates expansion of universe[11].

In 1929, Edwin Hubble discovered that galaxies are going away from us at higher speeds which are at farther distance and the speed is proportional to their distance. So the spectrum of light from more distant galaxies will have higher red shifts. From distant galaxies, light takes millions or even billions of years to reach us. This means what we are observing now is an image of universe of millions or billions of years old. In red shift, the spectrum is shifted from shorter wavelength to longer wavelength as the light from the galaxy travels and reaches us. The wavelength of light thus emitted increases due to expansion of the space itself over the years the light is travelling. If the wavelength of light received is found as doubled, then space must have expanded by a factor of two. This expansion factor is roughly proportional to the distance light travelled and that is the Hubble's discovery. Equivalently, it indicates how far back in time we are looking. This indicates that the universe was smaller and smaller as we go earlier and earlier. Tracing back in the timeline of the expansion of the universe, we see that the distance between galaxies become smaller while the density becomes higher. This continues until all matter is shrunk to a very small volume with an incredible density i.e. at the moment of the big bang[11].

From radioactive dating of isotopes of uranium, we can find out that the oldest isotopes were created (from nuclear reactions in supernovae) about 10 billion years ago. A study of evolution of stars reveals that the age of the oldest stars in our galaxy is about twelve billion years. This figure matches with the age calculated from the expansion of the universe observed. This observation suggests that the universe really started at finite time, supporting encouraging reason to believe in the big bang model of the universe.

In 1946, George Gamow, a student of Friedmann, proposed that the universe was very hot in the beginning because of the nuclear reaction that took place at the time of big bang i.e. at the point of singularity[1]. This process would have created helium and deuterium out of an initial sea of energetic protons and neutrons. The abundance of helium and existence of deuterium was the strong evidence that a extremely hot and violent explosion took place at the beginning and this was consistent with the proposed big bang model. In 1964, Penzias and Wilson were struggling to get rid of a constant background "noise" they were receiving along with the signals from the radio antenna. They tried to catch those pigeons that nested inside the horn-shaped antenna, and cleaned what they called "white dielectric material" produced by the pigeons on the antenna surface. After a year, even removing their apprehension of the problem due to pigeons inside the antenna, they still could not remove the background noise. They observed that this constant signal was precisely uniform in every direction and irrespective of directing the antenna to Sun or Milky way or any remote part of the sky. This meant that the signal was coming from far away places of our galaxy. Otherwise its omnidirectional nature would not have been seen. Physicists inferred that this must have been from the immense fireball of radiation at the big bang as Gamow predicted.

During the 1970s, this background radiation was observed by different groups of scientists at various wavelengths of microwave and infrared signals. It could be confirmed thereby that this background radiation had a blackbody spectrum with a characteristic temperature of about 3K.

Relatively recently in 1991, a satellite observatory called COBE (Cosmic Background Explorer) made a precise measurement of the background radiation from Earth's orbit and data so obtained perfectly matches with the spectrum of a theoretically ideal black body. Big bang model predicts that the universe at the beginning must have been very hot and full of particles and light,. In that condition, the particles were constantly bouncing in light, absorbing light and then re-emitting. Light emitted in such an environment would have displayed blackbody spectrum, and the characteristics of such spectrum remained intact while the light travelled through the expanding space[11].

Astronomers now know that the expansion of the universe during this time stretched out the wavelength of the radiation by more than 1000 fold. The afterglow of the big bang which is now being observed comes from the time when the universe was only about 500,000 years old. Considering the age of the universe this back ground radiation is as near as the event of big bang[11].

Hot Big Bang model

As per big bang theory, the gigantic explosion that occurred about 13.7 billion years ago caused the creation of universe. The subsequent evolution along with the expansion of the universe, the generation of light elements and the remnant radiation from the initial fireball, the formation of galaxies and other large-scale structures starting from one hundredth of a second after that incident up to the present day can be reliably described by this hot big bang model[1](different phases are described at Fig.2).

Big bang started as a point of nearly zero volume and enormous density. After big bang the universe was infinitely hot and as started stretching outwards in all directions the temperature of the radiation decreased. At this point, universe was not expanding within space but causing the expansion of space itself[1].

Planck Epoch

Very little is known about the period up to the first 10^{-43} seconds after the big bang known as Planck epoch, because our current physical theories cannot probe smaller timescales than this. However, it is imagined that all the four fundamental forces —, electromagnetism, strong and weak nuclear forces, and gravity — were unified into one force at this point, serving as one super-force, the super gravity. All matters, energy, space and time are supposed to have exploded out of singularity at big bang.

Grand Unification Epoch

The first spontaneous symmetry started breaking at 10^{-43} sec when the gravity force got separated from the combination of the other three forces-electromagnetism, strong nuclear force, weak nuclear force. The period from 10^{-43} to 10^{-35} seconds is called the Grand Unification epoch when the universe was smaller than a quark(one of the subatomic particles of matter) with very high temperature equivalent to 10^{27} K. At the end of this epoch, due to expansion and cooling, the particle energies would have gone down and there was phase transition resulting in breaking of symmetry between the forces and strong nuclear force got separated from weak nuclear force and electromagnetism around 10^{-35} sec. By then universe assumed a size of proton.

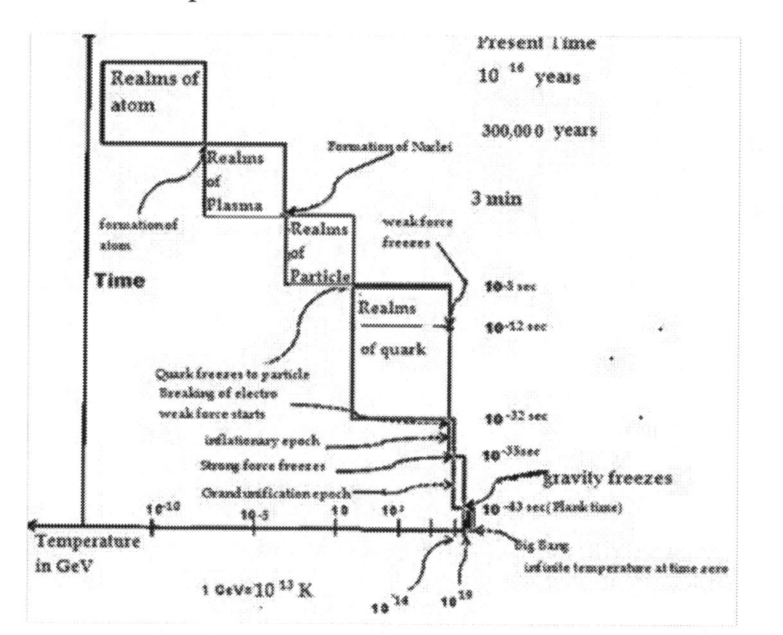

Fig.2.1 Time and temperature at different phases of big bang

Inflationary Epoch

Then came the Inflationary epoch that started 10^{-35} sec after the big bang and lasted up to 10^{-32} sec approximately and by this time the universe grew by a factor of at least 10^{26} or possibly more (Fig.2.1). It is anticipated that during this period the universe expanded at enormous speed i.e. more than 10^{30} times in only 10^{-32} sec approximately and the expansion during this period was more than that occurred during next 14 billion years. The universe became as big as a grape fruit at the end of inflationary epoch and the space expanded at speed many times higher than the speed of light.

Between 10^{-32} sec to 10^{-5} sec universe was consisting of energy in the form of photons. Because of extremely high energy density, these photons could not bind themselves into large stable particles. They were existing as collection of quarks.

One second after the big bang the temperature reduced to about 10 thousand millions degrees which is about thousand times more than the temperature at the centre of the sun. At this very temperature, it is assumed that the universe contained photons, electrons, neutrinos and their antiparticles along with some protons and neutrons [1].

With time and reduction in temperature due to expansion, the gas in the galaxies formed smaller clouds which started collapsing under gravity. With contraction, the temperature of the gas went on increasing until it became too hot to initiate nuclear reactions forming more helium converting hydrogen and the heat so released increased the pressure so that the clouds stopped contracting and remained in that state for long time. This happened in case of sun and other stars where this process is still continuing and hydrogen is getting converted into helium and the energy so radiated is providing heat and light (Fig.2.2).

With further expansion, the temperature further dropped and most of the electrons and antielectrons annihilated each other to produce more photons and only few electrons were left[1].

When the temperature of the universe had fallen to one thousand million degrees (as observed inside hottest star), protons and neutrons were not having adequate energy to overcome the strong nuclear force and therefore combined together to form nuclei of atoms of heavy hydrogen, deuterium (one proton and one neutron). These nuclei, combined with more protons and neutrons formed helium nuclei (two protons and two neutrons). In the process, a small quantity of a couple of heavier elements[1]might have been formed.

As the temperature further dropped and reached to about a few thousand degrees, the electrons and nuclei lose energy to overcome the electromagnetic attraction between them and started combining to form atoms[1].

In his book, 'the theory of everything', Hawking represented earth and other planets around the sun as the bodies formed by collection of some heavier elements out of the rotating gas cloud with a portion blown away at the time of formation of the sun, some five thousand million years ago.

Inflationary Theory

Though the big bang theory was widely accepted by the scientists, some problems with the theory could not be explained. When the big bang occurred, with the universe, hot radiation (energy in the form of waves or particles) released from the explosion, gradually expanded and cooled. This radiation, is called the cosmic microwave background, as mentioned earlier, appears as radio noise from all directions in space. This is the oldest light in the universe as we see now. This cosmic microwave background radiation was measured by the astronomers and it was found that its temperature was just below -270°C(3K). Calculations show that this is the correct temperature if the universe had expanded and cooled since the big bang.

But the radiation appeared as smooth, with no temperature fluctuations. The universe also would have expanded and cooled at a steady rate had the radiation cooled at a steady rate,. Then there could be no possibility of formation of planets and galaxies which were supposed to be clumped together due to gravity. This would

Big Bang	Era of grand unification	Universe was shaped	Formation of basic elements	Radiation era	Era of matter	Formation of stars and galaxies
	10^{-35} Sec	10^{-3} Sec	3 Sec	10000 Yrs	300000 Yrs	300 Million Yrs

Present day	Homosapien evolved	Dinoseurs are extinct	Appearence of mammals	Appearence of primitive animals	Formation of primordial life	Birth of Sun
	60000 Yrs	65 Million Yrs	200 Million Yrs	700 Million Yrs	3.8 Billion Yrs	5 Billion Yrs

Fig.2.2 Time line of Creation of Universe

have caused fluctuations in the temperature readings. Alan Guth, the American astronomer, proposed in 1980 one supplement to the big bang theory, called the inflationary theory which suggested that at first the universe expanded at an exponential expansion rate, i.e. at much faster rate than it does now. This astonishingly high expansion rate continued up to 10^{-32} second in the inflationary epoch i.e. after 10^{-35} second of the big bang and this concept of accelerated expansion allows for the formation of the stars and planets what we see in the universe today[12].

In general relativity, the average energy density of the universe decides the rate at which the universe expands and expansion is directly proportional to the density. This energy density includes the density of matter as mass of the matter is also a form of energy as per relativity. Other forms of energy are like electromagnetic radiation etc. In general, the rate of expansion reduces as the universe expands because the average density decreases[13].

Time for doubling the distance is considered proportional to one over the square root of the energy density and this is the relation between expansion and energy density. If there are 1000 galaxies in some region of space and all distances get doubled then the volume of space occupied by those galaxies will increase eight times. Since the galaxies have the total mass same as before, their density will decrease by eight times.

Thus if the mass of galaxies were the only form of energy in the universe then if distance get doubled due to expansion, the doubling time would increase by a factor of the square root of eight. Therefore universe whose energy exists fully due to mass will experience power law expansion[13(see Appendix B).

If there is a lot of light energy in the universe the doubling time increases faster than it would for a universe with only mass energy. For example, the energy density contained in light (which is a form of electromagnetic radiation) decreases faster than the energy density of mass. Every time the universe expands by a factor of two the energy density of light decreases not by eight times, but actually by 16^{-3}[13].

Therefore, if space is doubled in radius and the energy density in that region decreases by a factor anything other than eight, then the total energy would change. So, one problem appears with this fact that the total energy of the universe is apparently not conserved. The resolution of this problem is a somewhat delicate issue in general relativity and involves a kind of gravitational energy, associated with the expansion of the universe, which we cannot directly observe. As because the energy density that determines the expansion rate, does not include gravitational energy, the amount of energy that we can observe in the universe can change as the universe expands[13], though total energy including gravitational potential energy remain conserved,

Different kinds of fields react differently to the expansion of the universe. For example we have already mentioned the energy density of electromagnetic radiation decreases faster than that of ordinary matter.

During inflation, the expansion was exponential, or at least nearly exponential which means that the doubling time during inflation didn't change much as the universe expanded which means that the energy density in this phase did change very slowly. It is also known that inflation did last for a very short period. Therefore, to explain inflation, we would need to find a form of energy that changes very slowly for some period of time, but then begins decreasing rapidly[13] (see Appendix-B—Inflation).

The picture of universe provided by the big bang theory is mostly acceptable as of now. According to this theory our universe came into existence about 13.7 billion years ago in a violent explosion-a big bang. While the universe was expanding it was getting cooled. This cooling has caused the formation of different particles that subsequently forms matter. The particles like neutrons and protons combined to make primordial gases as and when big bang energy got sufficiently low, These gases due to anisotropies, created due to quantum uncertainty and other effects, started to concentrate in some regions of universe giving rise to formation of galaxies and stars. These details about creation of the universe is vindicated by both experimental discoveries and mathematical formulations.

There may be obvious question what happened before the big bang and what determined the initial conditions or properties of universe?

Primary building blocks of nature

The Greek scientists thought that the primary building blocks of the universe are particles in nature, the atomos, the word introduced by Democritus, the Greek philosopher to describe the ultimate, indestructible units of matter[3].

The twentieth century physics was struggling to find out the answers of a few questions[15][12]:

Why cannot we run away from a light beam and diminish its speed of approach?

We know that the light emerging from sun reaches earth after eight minutes. If sun explodes, does its gravitational impact on the earth's orbit will be felt after eight minutes?

Why the theories of physics dealing with macro-and micro-worlds of nature are found mutually incompatible?

The first question was raised because of an interesting conflict between Newton's law of motion and Maxwell's electromagnetic theory. Maxwell realized that the visible light was some sort of electromagnetic wave (see Appendix B) and his EM theory stated that the electromagnetic disturbances travel at a fixed and never changing speed which had been found to be equal to the speed of light. Newton's law of motion asserted that a person running away from light at a very high speed near to that of light would be able to reduce the speed of approach of the light. But Maxwell's theory disapproved this. Einstein, unknowingly entered into the picture and ultimately resolved the long pending paradox arising out of the above contradiction. He was trying to understand an imaginary situation. What will happen if we chase the light beam at speed of light ? As per Newton's law of motion, we shall catch up with the light wave and the light beam will appear as stationary. Light will stand still. Again according to Maxwell's theory, there cannot be anything like stationary light. Einstein through his special theory of relativity presented the solution to this puzzle which provided a fresh look to the idea of space and time. The theory states that space and time do not follow any set universal concept. Rather the form and appearance of space and time depend on one's state of motion. So, though for ordinary moving object thrown to us it is possible to run away from the object causing the speed of object slower in approaching us, for light beam this is not possible because as per Einstein's special theory of relativity, the speed of light is constant and speed of any object cannot be more than that of light. At great speed, time slows down and space is distorted[12](see Appendix-A).

Warping of space

Having said that, Einstein created another conflict with the observation of Newton, another stalwart of physics, that the influence of gravitational force would be transmitted instantaneously or at speed more than that of light over a vast distance of space. Einstein tried to resolve the conflict in his general theory of relativity offering a new concept of gravity and by showing that the gravity is transmitted through space warps. So far as fabric of space is concerned, it is flat in absence of any mass. When a large mass comes into picture, the space warps like a rubber membrane. But the resultant disturbances in the space will not be instantaneous. Rather it spreads out from the massive body like ripples in the pond when its water is disturbed and ultimately settle down in a warped space(Fig.2.3). The communication of this change in gravitational pull due to the massive body will be felt by other objects in space after some time i.e. the time taken by the impact to travel in space to reach the object from the location of the warping source. Einstein found that the effect will travel in the fabric of space of the universe at the speed of light precisely. So the explosion in the sun will change the mutual gravitational attraction, causes a distortion in the space-time fabric which travels at speed of light. So in earth, we see

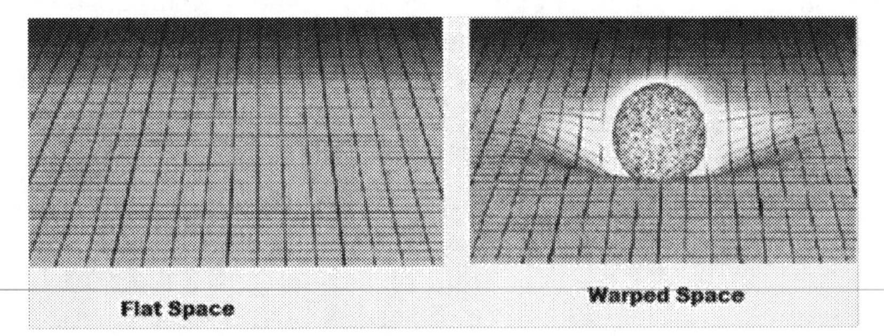

Fig.2.3(Fabric of space)

the explosion visually as well as feel the gravitational consequences at the same time. Thus the conflict is resolved. In his special theory of relativity Einstein stated that space and time are influenced by one's state of motion. In general theory of relativity he stated that space-time can warp and curve in presence of matter or energy. The space warps travel like ripple in the pond and gravity is transmitted at exactly the same speed of light[12].

Space-time is warped in presence of mass or energy. In relativity, we do not talk of space or time independently but talk of space-time.

Integration of Quantum Mechanics and General Theory of Relativity

The two pillars of twentieth century physics are Einstein's general theory of relativity and the quantum mechanics. The first one applies to the macro-world, the stars, galaxies etc. The other one helps to explore the micro-world, the atomic and subatomic structures of the world. These two remained mutually incompatible for long.

According to general relativity, the gravitational field manifests itself as curvature of space-time fabric represented by smooth Riemannian geometry(see Appendix-A). The quantum mechanics is armed with the uncertainty principle of Heisenberg that says that in micro-world, certain features cannot be known with absolute precision. As per this theory, at smaller and smaller regions of space, the amount of energy coupled in that space is known with less and less precision. There is a tremendous amount of throbbing, hot, kinetic energy associated in every little fragment of space and the smaller the fragment the more is the energy[15].

The uncertainty principle of quantum mechanics causes fluctuations of the gravitational field resulting in violent distortions in the space-time fabric. So the Riemannian geometry(see Appendix A) is no longer smooth and general relativity based on smooth geometry breaks down. Heisenberg explains that this incompatibility of the theory of general relativity in the micro world is due to the reason that when there is lot of energy in very small space, say around the size of a string (10^{-33} cm) the space becomes incredibly foamy and violently undulating. These undulations are so violent that they completely destroy Einstein's Geometrical Model of Space and the central principle of General Relativity. Therefore when the distance scales come down to the order of Planck length or less, the general theory of relativity gets invalidated.

The struggle for resolution of the conflicts between the two theories continued for half a century until the development of Super String Theory, which reconciles Quantum Mechanics with the General Theory of Relativity.

The compatibility between them is brought about by this new incarnation of quantum mechanics, according to which the particles of matter and fields are vibrating loops of strings.

As per string theory, the universe cannot be squeezed to a size shorter than the Planck length in any of its spatial dimension. In macro-scales, like that of galaxies and other celestial bodies, these miniscule kinetic undulations average out to zero and not visible. Only on microscopic scale, we become aware of the clamour that is going on and realize that its severity abolish effectiveness of Einstein's theory[15].

By Suzuki and Veneziano, string theory was accidentally discovered in 1968 as 'Dual Resonance Model', the properties of which were subsequently explained by Nambu in 1972 considering that the particles are nothing but vibrating strings. In 1980, Schwarz and Green set the superstring theory in motion and felt that the theory was mathematically consistent with the idea of 'Theory of Everything'. But after that, till early 1990s, not much progress could be achieved though the theory created a lot of hope in the minds of the physicists[12][15].

Conception of higher dimensions

Let us start the topic with a nice comment from Edward Witten of the Institute for Advanced Study in Princeton, New Jersey. He said," String theory is twenty-first century physics that fell accidentally into the twentieth century."[12]

For us it is not possible to imagine a world consisting of more than three spatial dimensions. Our brains perhaps are simply incapable of imagining additional dimensions. From the very childhood, the concept of three dimensions of space has deeply embedded in our brain. The concept was further extended by Einstein who enlightened the world with the conception of the fourth dimension- the time. But today the scientists are harnessing the possibilities of higher dimensions, the spatial dimensions beyond conventional four dimensions of space and time. As mentioned earlier, the big bang theory very often faces the questions, why the universe exploded or whether before big bang something similar happened. Scientists and theologians were at a loss while trying to find out the reason behind origin and nature of the big bang. The superstring theory has attempted to explain the puzzle by conceiving the idea of a ten-dimensional universe prior to big bang.

String theorists are of the opinion that additional dimensions exist and in fact, superstring theory requires at least ten dimensions for the universe.

We know that the string like DNA molecule is the essential feature of life on earth and it contains complex information and coding of life itself. Similarly, the strings are supposed to build up the subatomic matter and ingredient of life having large amount of data stored in them in a relatively simple and reproducible structure.

Nature's exposition becomes simpler in higher dimensions[2]

The great debate prevailed in the nineteenth century regarding propagation of light from stars to earth travelling trillions of trillions miles through vacuum when light could be verified experimentally as wave. But how a wave can propagate through vacuum when there is no medium? To explain this the physicists put up a conjecture that a substance called 'aether' filled up the vacuum and it acted as the medium for the light wave. Subsequently existence of aether could not be established experimentally[2].

Polish mathematician Kaluza, in 1919 proposed that the general relativity and electromagnetic theory could be amalgamated with the inception of a fourth spatial dimension. The Swedish mathematician Klein added that space consisted of three spatial dimensions with which we were familiar and a curled-up dimensions found deep within the three spatial dimensions and could be thought of as a circle. But Kaluza and Klein's curled-up dimension could not unite general relativity and electromagnetic theory as originally anticipated, but later on string theorists found that the idea could be made useful.

Kaluza-Klein theory, a theory of gravity, stated that light wave was a vibration of fifth dimension and light wave travels through vacuum, as the vacuum existing in four dimension of space and one dimension of time, itself vibrates. However due to technical problems in describing sub-atomic particles, the theory could not be made useful for about fifty years[2].

To build theories which are compatible with quantum mechanics as well as special theory of relativity, a model known as Standard model was designed on the basis of Quantum Field Theory(QFT) and according to this model, a particle is considered to be a point moving through space. But in order to explain the interactions taking place in the nature, a particle must have more degrees of freedom which can be characterised not only by position and motion but also by mass, electric charge, spin or colour (it is the charge associated with strong force interaction). Though the standard model could unify three of the four fundamental forces, i.e. electromagnetic force, weak force and strong force, it could not integrate, by its postulations, the gravity which is the fourth type of interaction in nature and nicely described by Einstein's general relativity.

Superstring theory

It was super-gravity theory and particularly superstring theory which eliminated the difficulties of Kaluza-Klein theory and a new theory called theory of Hyperspace or the theory of higher dimension evolved and a great revolution took place in the concepts of the modern theoretical physics for unveiling the secrets of the nature and creation of the universe and the theory subsequently served a very important role in spearheading the attempts to explore the theory of everything[2] (see Appendix-B for string theory).

CHAPTER III

CREATION OF THE UNIVERSE–THE CONCEPTS

The theological thought processes

Universe as per Hindu philosophy

We have seen that there are a number of theories regarding origin of the universe but now big bang model is the most acceptable theory by the scientists. It is very startling to see that the sages of Upanisads knew about the cosmic events through their mystic insights.

The Vedas was written in the Vedic Sanskrit form some time around 2000-1000 BC and propagated the thoughts of the spiritually inspired thinkers about origin the divine universe and how they felt oneness with it. The Upanisads are the most recent compilation of Vedas dated back around 600 BC. These are one of the most concise and thought provoking expositions of the spiritual experiences which were at the base of all the great religions. Even though nearly three thousand years old, discourse of the teachings are as relevant today as they were first preached by the ancient sages of India.

It was described in the Chhandogya Upanishad:

sad eva, saumya, idam agra asId ekam evadvitiyam, taddhaika ahuh, asad evedam agra asId ekam evadvitiyam, tasmad asatah sat jayata

[My dear boy, in the very beginning, a single Reality was existing. It was only one reality and no other. It was one and only one. It was matchless and having no external ramification. Nothing could be equated to it or be separated. There are some opinion that in the beginning, there was nothingness alone and everything was created out of nothing.][39]

Kutas tu khalu, saumya, evam syat, iti hovaca, katham, asatah saj jayeteti, sat tu eva, saumya, idam agra asid ekam evadvitiyam.

[But my dear boy, Can Being be evolved from 'Non-Being'? You see that something is produced from something. How nothing produces that something? So you should conceive that Being alone was there and not Non-Being. This Being is non-dual Being. It is not an individual or Being. It is not a personal Being. It is unimaginable.][39]

Tadaikshata bahu syam prajayeyeti tattejo' srijata

["He, the Brahman, the absolute reality, desired to be many and there after grow. He made the universe out of himself and then he entered into every being. Thereby in all the things he is the delicate spirit, the absolute truth and the eternal Self.][He, the Brahman, the absolute reality, desired to be many and there after grow. He made the universe out of himself and then he entered into every being. Thereby in all the things he is the delicate spirit, the absolute truth and the eternal Self.][108]

Here creation of Akasa by Brahman had not been mentioned. But there was a Shruti text that indicated that Akasa was created.

Tasmad va etasmadatmana akasah sambhutah

[From the Self (Brahman), Akasa was created, from Akasa the air, from air the fire, from fire the water, from water the earth][108]

The Vedic thinkers tried to understand the beginning, growth and end of the universe. They tried to learn the absolute which regulates not only our life but the whole universe. To reach the people at large, they took the help of Brahmanas, Puranas to explain the things using mythology. They called the universe 'Brahmanda'.

Vedic philosophical thought was expressed through Nasadeeyasukta which said,

Nasadasinno sadasittadanim nasidrajo no
Vyoma paro yat |
Kimavarivah kuha kasya sharmannambhah kimasid gahanam gabhiram ||

[There was neither existence (*sat*) nor non-existence (*asat*). There was no sky, no space too. Then what is the reason of that arousal ? From where? what were the effects? Was there infinitely deep water? (something like *gahanagambhira)][26]*

Namrtyurasidamrtham na tarhi na ratrya ahna asit praketah |
Anidavatam svadhaya tadekam tasmaddhananna parah kim chanasa||

[There was no death, no timelessness. There was no Night, no Day. That absolute omnipresence was breathless, windless. There was nothing else].[26]

tama asittamasa gur'magre'praketam
salilam sarvama idam |
tuchchhyenabhvapihitam
yadasittapasastanmahinajayataika||

[Darkness was covered by darkness in the beginning, no distinction all water, the life force originated out of emptiness, that ONE appeared through the power of heat.][26]

Kamastadagre samavartatadhi manaso retah prathamam yadasit |

[Desire emerged out of that ONE in the beginning, That was the first kernel of mind.][26]

Aitareya Upanisad says,

Atma va Idamek Evagrya Aasit / Nanyat Kinchan Misat /
Sa Ikshat Lokannu Srija Iti //

[Before creation of the universe everything was dissolved within the Supreme Personality of the absolute Being (Brahman), everything was confined within Him, no isolation. The Jivas (individual souls, tatastha sakti) and even the external energy Maya was also within Him. So that Lord then desired to manifest everything of the worlds through me][27]

The perception described through the above hymns about the creation of the universe represents a shadow of the big bang theory.

As per cosmology, the universe suddenly emerged from a point source of infinite energy, called space-time singularity. In Vedanta also the same realization was expressed by saying that the absolute reality, consciousness and bliss of the universe emerged from a single source.

The Isha Upanisad says:

Om Puurnnam-Adah Puurnnam-Idam Puurnnaat-Purnnam-Udacyate Puurnnashya Puurnnam-Aadaaya Puurnnam-Eva-Avashissyate

[Om. That Supreme Brahman is infinite, and this adapted Brahman is also infinite. The infinite derived from infinite. Then through knowledge, realizing the limitless of infinite, it remains as the Infinite alone.][28]

Advaita Philosophy

As already said, the Upanishads describe the creation, maintenance and end of the universe. But according to Advaita, creation is not real, it is only a camouflaged image of Brahman, which is the only real in the absolute meaning. The universe, which is a manifestation of *maayaa,* is called *anirvachaniiya.* It cannot be described either as real or as unreal. It has reality to be realized by self only. Upanishads have described creation of the universe only to bring out the truth that Brahman is the cause and alone is real. The universe is the effect, and has no independent existence other than Brahman.

According to Advaita, the effect is not an actual transformation of the cause. Brahman is immutable and there can be no transformation of it. It only serves as the substratum (*adhishThaana)* for the appearance of the universe like a rope appearing as a snake due to illusion,

Therefore the universe appears due to *avidyaa,* or nescience, and conceals Brahman by its veiling power (*aavaraNas'akti).* The universe is therefore said to be only a *vivarta,* or apparent transformation of Brahman. That means that the universe is real for all those who are still ignorant about Brahman. It loses its reality only when Brahman is realized as the only reality and as identical with one's own self, or, in other words, when identification with the body-mind complex (subject-object duality) completely disappears.

As per the Aitereya Upanisad there was only one Atman at time before the universe came into being. There was nothing moving. Then Atman thought of producing the universe. The sage Aitereya described that "He, the absolute soul, thought of creating the worlds". Here the sage was convinced that the creation of the Universe was thoughtful and was not an accident. In fact, Sir Lowell showed that the primitive fireball that exploded and produced this universe at the time of big bang was with precision in timing. He added further that had there been a mistake of a fraction of a second there would not have any Hydrogen formed and then this type of universe could not have been produced. Accordingly the idea was being made that there was some thought or knowledge based on which the Big Bang took place. This view was given also by the sage Vasishtha long long ago, the record of which was available in Mahabharata.

With Sir Lowell's observation, a basic question may be raised how the big bang operation was conducted with so much precision and who did it? In order to avoid this question we may take shelter of the steady state theory of the universe but the same has already been rejected by most of the cosmologists due to other stellar observations. Also, as time is beginning from the instance of space-time singularity as per big bang theory, the answer to the above question is also not available. But the Vedic philosophy kept the answer ready even before thousands of years.

In Brihadaranyaka Upanisad the answers related to space, time and universe were searched as we see from the conversation between Gargi and Yajnavalkya. Gargi started to ask the questions to know about the celestial hierarchy[9].

Gargi: *yajnavalkya, iti hovaca, yad idam sarvam apsv otam ca prota ca, kasmin nu khalv apa otas caprotas ceti?*

['OYajnavalkya, if everything is otapoto(warp and woof) with water (jal), with what water is otaproto?]

Yajnavalkya: *vayau, gargi, iti*

['Gargi, with Air (vayu).]
Gargi: *kasmin nu khalu vayur, otas ca protas ceti ?*

[With what is Air otaproto?]
Yajnavalkya: *antariksa-lokesu, gargi, iti.*
[Gargi, With Antariksha.]

Gargi: *kasmin nu khalu antariksa-loka otas ca protas ceti.?*

[With what is Antariksha otaproto?]

Yajnavalkya: *gandharva-lokesu, gargi, iti.*

[Gargi, withGandharvaloka]

And the questions continued in succession as Jal-Vayu-Gandharvaloka-Adityaloka-Chandraloka-Nakshatraloka-Devaloka-Indraloka-Prajapatiloka-Brahmaloka.

Yajnavalkya got angry when Gargi still continued to ask 'where is Brahma loka'? Yet Gargi got some conception of space or the universe. Being on earth, it is possible to realize that every thing is deeply associated with water and most part of earth is covered with water. Vayu or air is the apamsakha (friend of water or jal). Vayu is spread in antariksha, the atmospheric world, the sky and is source of relation with other planets. Gandharva is the rays of sun and therefore Gandarbhaloka is where the sun's rays cross. *Adityaloka* is the region of the sun. This solar region is located in Chandra loka. Here by Chandra we do not mean the physical moon that we see, but the original condition of the planetary substance from which all the stellar regions can be said to have come out as effects from the cause. They are called *Chandra-loka* because they are eternal in nature and not solid masses or globes shining like the stellar region. *Chandraloka* is located in *Nakshatraloka,* the region of stars. This located in Deva loka, the region which causes the appearance of stars. The astronomical appearance of these cosmic entities are in this celestial region called the celestial paradise. Then the *Devaloka* is interlaced like warp and woof with *Indraloka.* The *Indra-loka,* the world of the ruler of the heavenly bodies, that is superior to the location of the ordinary celestials. Indra is the Surya and the *Indra-loka* is the source and is prior to the celestial paradise. It is subsequent to *Prajapati-loka.* Here *Prajapati-loka* is synonymous to *Virat* himself and *Virat* is the Supreme Power of all the worlds, *Indra-loka* and everything downwards. Beyond *Prajapati-loka*it is *Brahma loka* which is the world of *Hiranyagarbha.* It is the source of all. As beads are strung on a thread, everything that is created is strung in this Ultimate Being. Everything is included in it in one form or other. It is also called *Sutra-Atmanor* the *Brahma* which is the source and cause of everything[9].

When Gargi asked details of Brahma loka, Yajnavalkya wanted to stop her by saying that she was crossing the limit because she was asking the cause of the cause of everything or cause of the supreme cause, the ultimate reality. Actually, Yajnavalkya perhaps wanted to tell that unless one transcended the state of mind to such extent to overcome the sense of subject-object duality and achieve self realization it would not be possible to understand the *Brahma loka.* Also once got dissolved in ultimate reality, the Brahman, one loses one's identity and become the reality, the unknowable itself. Thereafter he cannot describe the reality once comes out of that state of mind[24].

The word *'Brahman'* is derived from the Sanskrit root 'brh' which means to grow big without limit and indirectly indicates an explosion. When Indian sages realized the Absolute through transcendental meditation, they felt the need for an adequate symbol to communicate the inexplicable Brahman or Absolute Truth. Their theological investigations led them to believe that at the beginning there was a Sphota (Big Bang!) which made a sound. This sound (shabda or word) was Nada Brahman or Shabda Brahman. Everything like space, time, consciousness, energy, all came out of this Shabda Brahman, AUM[29] over the years. No particular letter, a consonant or a vowel of the alphabet, could amply express this sound. AUM is considered the unity of all sounds to which all matter and energy are reduced to their primordial form. The primordial sound of creation, the 'AUM', symbolizes the manifesting vibration or sound of the big bang. The Shabda Brahma or Brahma is considered as the impersonal absolute of pure timeless existence including everything before big bang and also the whole manifested universe after big bang. It is eternal, genderless, omnipotent, omnipresent, omniscient and indescribable.

Samkhya Philosophy

It will also be appropriate to mention here about the Samkhya system of the Hindu philosophy. Samkhya is one of the most prominent and one of the oldest of Indian philosophies. An eminent, great sage Kapila founded the Samkhya School. Samkhya means number and it also means perfect knowledge. So it is system of perfect knowledge.

In this system the creation out of nothing (ex nihilo) has been negated and is thought that the effect pre-exists in the cause. It says that there cannot be a new creation but the creation is nothing but the manifestation of what was already present in the cause in a potential form. As per Samkhya there is no creation of the universe by a god, but rather a continued evolution of the universe in which matter and energy interact.

Samkhya is dualistic realism i.e. *Dvaita*. It is dualistic because it advocates two ultimate realities and describes the principle that cosmic consciousness is composed of two primary energies—the masculine energy (subject), called as *Purusha*, the self or the spirit and the feminine energy (object) called as *Prakriti*, the matter. It is the union, or synergy between *Purusha* and *Prakriti* upon which the entire universe is created. Their eternal love and desire for one another being stimulated by *prana*, concede *Purusha* and *Prakriti* to operate in perfect mirrored unison. *Purusha* is the supreme independent consciousness, and *prakriti* is the primordial substance behind the world and is the material cause of the universe. *Prakriti* is the main and ultimate cause of all gross and subtle objects. *Prakriti* is unintelligible, non-self and devoid of consciousness. It gets greatly influenced by the *Purusha*, the self. we have *mahat* with union of *Purusha* and Prakriti and then that itself is called *ahamkara* when it is assertive in a cosmological sense. Through the cosmic union of pure awareness (Purusha) and primordial matter (Prakriti), Mahat or universal intelligence, is born with Prana (breath or life-force) and helping to carry on the process. Mahat is absolute intelligence which often appears to us in the form of divine grace and intuition. In order for Purusha to experience itself, separation has to occur. Such separation happens via the Ahamkara (creaor of 'I'). The Ahamkara, or ego, is the force which provides us with our individuality and self-identification, forming the basis of the Law of Karma. It is individualized awareness focused on identifying with the limited world[30]. Beyond *Purusha* and *Prakriti* it is Brahman, which contains no duality, no subject-object, no witness or witnessed.

In evolution, *Prakriti* is transformed and disintegrated into multiplicity of objects and evolution is followed by dissolution when all the worldly objects gets united and go back to the primeval matter[25].

According to Saṃkhya philosophy, the root causes from which the entire universe is originated *are Sattva, Rajas* and *Tamas*. When there is mutual equilibrium in their states, they are called to be in *Procrit*, the homogeneous state and when their balance is disturbed, they are said to be in *Vikṛti*, the state of heterogeneity.

These three *guṇas* are said to be the ultimate cause of all creation. *Sattva* is weightlessness, light and pleasure, purity and fineness ; *rajas* is motion or activity and pain ; and *tamas* is heaviness, inertia and inaction, negligence and indifference.

The *guṇas* constantly change their predominance over one another. The *gunas* are always changing, imposing dynamism to *Prakriti*. Yet, the three *gunas* maintain a balance. The flow of cosmic and individual life continues through their constant interactions. They are essentially different forms but interrelated with one another and different *guṇas* co-operate with one another to produce the objects of the world[25].

Prakṛti requires *Puruṣa* in order to be known; *Puruṣa* requires *Prakṛti* in order to distinguish itself from *Prakṛti*. The interaction of *Puruṣa* and *Prakṛti* causes a disorder in the equilibrium of the *guṇas*. Being inflamed, Rajas causes the other two *guṇas* to vibrate; the vibration releases tremendous energy within *Prakṛti* to destabilize the equilibrium and from this the universe is manifested. This process of manifestation goes through 24 stages, and begins with the infusion of *Puruṣa* into *Prakṛti*.[25]

Let us now divert our attention to other philosophies of the world and the conception of creation of universe thereof.

Buddhism about Creation of universe

While commenting on the origin of the universe Bertrand Russell said, "there is no reason to suppose that the world had a beginning at all. The idea that things must have a beginning is really due to the poverty of our thoughts"[31]. In fact this is similar to that depicted by Samkhya philosophy which explains how the universe is created and describes that the beginning of this world and of life is inconceivable since they have neither beginning nor end.

In an attempt to explain the universe, its origin, and human's role in this world, the idea of God was found entirely unnecessary to the Buddhist thinkers. Through the centuries, Buddhist philosophers have tried to bolster arguments to disprove the doctrine of a creator god.

Buddha described the origin of the Universe and life in the Agganna Sutta about twenty five hundred years ago. According to Buddha's proposition the universe being created and destroyed repeatedly over a period of millions of years and has reached the present form. Buddhism never claimed that the universe comprising of the galaxies, other celestial bodies, water, wind, days and nights were created by a powerful god or by a Buddha. According to Buddhism, world systems always appear and disappear in the universe and this is a continuous process[31][34]. In the eyes of the Buddha, the world is nothing but Samsara—the cycle of births and deaths continuously repeated. The beginning of the universe and the end of it is cycling within this Samsara. Since the matter and energies are relative and intermingled, it is not wise to distinguish anything as the beginning[31]. The Buddha did not waste His time on this issue. The reason for His reticence was that this issue has no religious value for gaining spiritual acumen. To provide the clarification of the origin of the universe is not the concern of religion. Such conjecturing is not necessary for living a righteous way of life and for shaping our future life[35]. In fact, Buddha, is supposed to have said in the *Acintita Sutta*, "Conjecture about the world is an unconjecturable that is not to be conjectured about, that would bring madness & vexation to anyone who conjectured about it."[32].

However, Buddha presented a model of cosmology wherein the universe is described as expanding and contracting over extremely long periods of time and this description appears to be consistent with the expanding universe model and Big Bang[34].

The Buddha explained that the universe expands outward, reaches a threshold point, and then reverts its motion gets reverted toward a singularity resulting in its destruction. This process is repeated infinitely[34].

Buddha, in Agganna Sutta is found to speaking to the monk Vasettha, a former, Brahamin that "Now there comes a time,

Vasettha, when after a long period of time this world expands. By world here the Buddha is meaning the Universe. The Buddha says "Space is endless, and there are countless numbers of worlds." This indicates that the universe has no limit, and has an endless number of stars and planets[34].

Creation of universe as per Islamic Philosophy

The scholars dealing with subject are of the opinion that the creation of the universe is a subject that is given great importance in the Qur'an. The huge and varied amount of information contained in the Quran about almost every stage and aspect of the creation continues to astound scientists today because of its very accurate agreement with current knowledge about the universe. The book was written in the 7th century and contains lot of scientific information that actually came into light 14 centuries later?[33] Much of this information has been discovered only in the last sixty years! Neutral and unbiased observers do consider this to be valid evidence that such a book could never have been the product of any human being for the simple reason that, at that time, no human possessed such knowledge.

According to The Qur'an, the life containing universe is solely created by the Almighty God—Allah—and He is the Supreme Ruler of the universe[37]. The Quran also describe that this universe—Earth and Heavens—has definite life period and has some definite objectives like other creations.

Qur'an describes that the creation of the universe is by Allah's will with a single command: "Be!" Several verses in the Qur'an highlight Allah's power of creation from nothingness. When He decrees a thing, He says "Be," and it is[33].

In Qur'an, God says: "We have built the heaven with mighty force and We have undoubtedly expanded it."[33] It appears to have given clue about the expanding universe. Qur'an also says: "He raised its (the heaven's) height and brought about the synergy in its constituents."[33] It appears to mean that the space is created and the laws to govern the universe i.e. gravitational force, interaction of primeval particles for creation of matter and time is instituted.

One of the verses from Qur'an says: "The unbelievers do not see that the skies (space) and the earth (matter) were joined together (as one unit of creation) and we shredded them."[33] In fact this resembles the happening at the beginning of the creation of the universe which resulted from an extremely dense singularity. The creation of the universe resulted matter, space and time that are intimately linked together. Initially the matter and space were joined as one and then were separated in the explosion.

Another verse says: "Then He took hold of the sky when it was smoke."[33] Actually it has been scientifically expressed that the whole event of big bang took place at a very high temperature; it was a hot Big Bang. The universe in its very early stages was, thus, in the form of hot gases.

In this way, there are number of verses can be found in Qur'an those can be interpreted as correct descriptions of the genesis, subsequent formations of matter and the life of the stars etc. of the universe.

Islamic thought categorizes the entire cosmos into two domains: the Unseen Universe , which is not conceivable to mankind in general, has characteristics not known to us, and includes Allah (metaphorically), angels, Paradise, Hell, seven heavens, and *Al-Arsh* (the Divine Throne). The Observable Universe is perceptible through the five senses (possibly enhanced by means of instruments). The Qur'an says: "He, Who is the knower of the Unseen and the Observed is Allah, the only God. [37].

It is described that the human beings were bestowed with special position in the universe created by Allah. He then left the world to function under the laws. He had authorized for it. He observes how people take the privilege of enjoying the abundance provided for them by Him and the system is gifted to us and exist, with little intrusion, for a certain length of time. At the end of this time, mankind is brought in front of Allah for Judgment Day to be punished or rewarded for his deeds in His universe.

Jainism and creation of universe

According to Jains, the Loka or universe is ever present in varying forms with no beginning or end. According to Jain doctrine, the universe and its constituents—soul, matter, space, time, and principles of motion have always existed (as depicted in steady state cosmological model)[40]. All the constituents and actions are governed by laws of nature. It is not possible to produce matter out of nothing and hence the sum total of matter in the universe remains the same (similar to law of conservation of mass).[41] Jain text claims that the universe consists of *Jiva*(life force or souls or living substance), and *Ajiva*(non-living substance). Similarly, all living beings are characterised by the unique souls and cannot be created. They are existing since beginning.

The Jain theory of causation holds that a cause and its effect are always identical in nature and hence a conscious and immaterial entity like God cannot create a material entity like the universe. In fact Jainism does not support belief in a creator deity. Furthermore, according to the Jain philosophy, any soul who can come out of the bindings of its karmas and desires, achieves emancipation / Nirvana. A soul who liberates itself from all its passions and desires has no desire to stay tuned to the working of the universe. Desire for rewards and freeing oneself from sufferings are not the work of a divine being, but supposed to be an eventuality of innate moral order in the cosmos; a self-regulating mechanism whereby the individual reaps the fruits of his own actions through the workings of the karmas[42].

Through the ages, Jain philosophers have adamantly rejected and opposed the concept of creator and omnipotent God and this has resulted in Jainism being labelled as *nastikadarsana* or atheist philosophy by the rival religious philosophies. The idea of omnipresent universe and absence of omnipotent God with divine elegance runs strongly in all the philosophical dimensions of Jainism, including its moral code of conduct. Jainism asserts that a spiritual and righteous life is possible without assimilating the idea of a creator god.

Jain scriptures do not accept God as the creator of universe. Jainism offers an detailed cosmological picture including divine beings/ Devas. These divine beings are not viewed as creators, they are adaptable to suffering and change like all other living beings, and must eventually die. Jainism says that if one's soul is freed from Karma and he attains the state of enlightenment or Nirvana he achieves the state godliness and known as Thirthankara. Thus, Mahavira was such a God/Tirthankara[42].

What is said in Chinese Philosophy

The two most influential spiritual leaders of China, Confucius and Lao-tzu taught the philosophies that they practiced in 6th century BC which are known as Confucianism and Taoism which existed simultaneously in China, attracting innumerable numbers of followers over the past 2,500 years[45].

According to Taoism, the entire universe and everything in it flows with a mystic, inexplicable force called the Tao. The Tao has lot of diverse meanings. It is the name that describes absolute reality. The Tao also explains the energy that drives the universe and the wonder of individual's nature. The Taoists have the conviction that everything is one despite all appearances. In its original cosmic sense, the Tao is the absolute, inconceivable reality and effectively it is the equivalent of the Brahman of Hindu philosophy and the Dharmakaya of Buddhism. It differs from these Indian concepts, in the sense that the Chinese thinkers viewed the intrinsically dynamic quality of Tao as the essence of the universe. The Tao is the cosmic phenomenon involving all the things; the world is seen a continuous flow and change[45].

Later on a Taoist religion evolved which was different from the philosophy of Taoism as described. This religious Taoism held some of the same beliefs like Tao philosophy, it was mainly for worship of many gods and ancestors, a practice which started during the Shang dynasty. Other religious practices included the refinement of bodily energy called "chi," the creation of a system of morals, and use of alchemy in achieving immortality. This Tao religion became popular after its adoption by China as the state religion in 440 AD and continues to be practiced even to the present-day[45].

Confucius born in 551 BC was a politician, musician, and philosopher and propounded another set of philosophical thoughts. He wandered throughout China, first as a government employee and later as a political advisor to the rulers of the Chou dynasty. Confucianism became the social philosophy of China from the Han dynasty in 202 BC until the end of dynastic rule in 1911.[45] It was the philosophy of social understandings, application of common sense and use of knowledge. It provided Chinese society with a strict conventions of social etiquette and firm base of education. Its main intention was to form an ethical basis for the traditional Chinese family system with its complex structure and its rituals of ancestor worship.

Though two different sets of popular philosophies, Taoism and Confucianism have lived together in China for well over 2,000 years[45], Confucianism affiliates social matters, while Taoism involves itself with the search for meaning[45].

Christian philosophy and creation of universe

About 2500 years ago or more Moses, David, Isaiah, Jeremiah, and other Bible's prophets and apostles stated explicitly and repeatedly the two most fundamental properties of the big bang, a transcendent cosmic beginning at a finite time before and a universe undergoing a general, continual expansion. In Isaiah's book both the properties were propounded, "This is what the Lord says—He who created the heavens and stretched them out."[46]

In Hebrew scriptures(old testament) the word "created" translated as *'bara'* which means "bringing something into existence new, that did not exist before." It described this supernatural creation as "the universe that we humans can measure and detect is made out of that which we cannot measure or perceive." The massage of Isaiah and others was that God alone was the messenger for the universe's existence. [46]. The characteristic of the universe stated more frequently than any other in the Bible by repeating the word "stretched out.".

Bible describes the stretching out of the heavens as both "finished" and "on going". This simultaneously finished and on going aspect of cosmic stretching is identical to the big bang concept of expansion of universe. According to the big bang theory, at the creation event all the laws, constants, and equations of physics are instantly created, designed, and finished in such a way to guarantee an on going, continual expansion of the universe and that too at exactly such right rates with respect to time that physical life was possible[46].

This biblical claim for simultaneously finished and on going acts of creation, incidentally, was not limited to just the universe's expansion. Isaiah and Zechariah commented about laying of Earth's foundations by God. This is consistent with the geophysical discovery that certain long-lived radiometric elements were placed into the earth's crust a little more than four billion years ago in just the right quantities so as to guarantee the continual building of continents[46].

Romans 8, a popular chapter of Bible has recorded that the entire creation has been subjected to the law of decay (the second law of thermodynamics) and indirectly argues for a big bang universe by stating that gravitation, thermodynamics, electromagnetism have universally operated throughout the universe since the cosmic creation event itself. Genesis and many other biblical chapters establish that the cosmos was much hotter in the past and the stars have existed since the early times of creation. It tries to establish that even the slightest changes in either the laws of gravity or electromagnetism would make stars impossible. In fact, in cosmic environment, the orbits of planets around the stars and in atom model, the electron orbits around the nuclei of atoms would not have been stable if the gravity, electromagnetism, and thermodynamics are not working as per the pre-set laws imposed at the time of big bang[46].

It is the traditional Christian theism which states that creation *ex nihilo* is miraculous—something which the laws of nature cannot explain. But a number of theists take the past-singular nature of cosmological models as confirmation of the claim that God created the universe *ex nihilo.* Again not all theists are committed to the claim that the universe is finitely old. Some contemporary theologians like Peacocke and Barbour even claim that the doctrine of the "creation" of the universe is best interpreted as one of the timeless dependence of universe on God, and that such dependence does not demand a temporal creation event[47].

The Christian theism is committed to a belief in a finitely old universe primarily because of its commitment to the accuracy of Biblical accounts of creation. But many theists do not agree to imagine the finiteness up to the extent of billion years as claimed by the promoters of the big bang theory. As per Bible, the age of the present universe is much less than the Astronomers' claim. So, if a theist comes to believe the big-bang account of the origin of the universe and therefore doubts the literal Biblical account, then he will hurt the basis of the theism i.e. the literal accuracy of biblical account, only for believing that the universe is finitely old.

CHAPTER IV

QUEST FOR GOD—THE SUPERNATURAL POWER

About 40,000 years ago, the human civilization marched a step forward with Upper Paleolithic Revolution marked with the appearance of modern Homo Sapiens with advanced social and cultural activities. It was the neurological transformation taking place in the brains which augmented the capacity of humans to generate higher level of sense in realizing the necessity of better life to live better[49]. This had driven them to formulate some type of social bindings. To establish that in the minds of the people, the idea of some supernatural power was conceived so that every body while performing in the social environment, keep the same thing in minds and try to adhere to the social requirements either in fear or to remain protected under the shelter of that supernatural power. This gave birth of the organized religion later on, characteristically different in different parts of the world depending on the geographical locations, compulsions contributed by the nature and peculiarity of the particular groups.

In the subsequent Neolithic Revolution, humans shifted from their nomadic life style and focused on agriculture. The settlements came up living along with cattle and plants. The food habits got changed and the individual efforts to gather foods changed to group farming. The population explosion resulted due to life span expanded and higher birth rate[49]. To maintain the social order, organized religion was established to inculcate same belief and commonalities in the group.

Religions and Concepts of God

It is very peculiar that irrespective of the locations, languages, constitution of the groups, different living conditions, different ways of life and even without the communication links between the groups, there was an universal overwhelming passion for the quest for some super natural power associated with some entity of God when actually there was no practical direct evidence of such entity could be found out. The majority of the population of the world have belief in the existence of God.

The primary argument of the remaining non believers in favour of their disagreement with existence of God is that God cannot be seen. This impression cannot stand because the believers immediately substantiate the ground of their disbelief by the counter statement that 'cannot be seen' is not true in absolute terms. Their argument is that it is important—how you are seeing. In a room, we do not see any thing in open eyes. But we know that the room is filled with air. It is also filled with radio waves which are beyond the range of natural human sensing capability. If you have the necessary detectors to detect the air or radio waves in the room, you will definitely be able to establish the existence.

There is another strong reason of belief in God is that the universe, even though continually reforming itself and possess a very complex but flawless running, is absolutely balanced at macro as well as micro level, obeying the laws of nature. There is a word 'Anthropic coincidences' which relates to the 'precisions' in the various parameters we observe in the running process of the universe that have caused possibility of life in earth. For example, the elements on which life depends such as Carbon, nitrogen, oxygen and Iron, all are the products of nuclear fusion reactions within the stars. If the initial condition of the Big Bang was different, nuclear fusion and supernova which distributed these elements would not have been possible. Moreover, there are four fundamental forces-weak and strong nuclear interactions, gravitation, electromagnetism which govern the formation of the elements. If the relative strengths of these forces become different, the resultant universe would have been different and the possibility of lives in it would have been far from reality. Why this has happened that too with so much precision is not known and it is also certain that according to laws of nature with which we are equipped now, any small changes in any of these key parameters would have resulted in a grossly different universe. This apparently mysterious incidence provokes the thoughts to imagine some super natural entity. It was felt that behind such accurately calculated, well planned creation, there must be some creator, called as God by the believers. By such groups of believers having beliefs in existence of God or ultimate reality, organized religions were formed.

The earliest documented organized religion is Polytheistic religion dictating belief in more than one God. In ancient Greece, ancient Egypt and ancient India, polytheism emerged retaining some of the characteristics of the early religion prevailing at that time. The belief in the supernatural power was depicted through imagination of god in form of animal, human and some time combining animal/human features which was tried to be displayed in cave painting, rock carving and statues even prior to establishment of the polytheism[49].

God in Greek religion had human forms, male and female. The Egyptian gods had also the same form These gods had the abilities to take the form of animal or trees. In Hinduism also there were so many gods and even in the animal form like Ganesha.[49]

In polytheism, different gods had different responsibilities to run the world. So people used to get involved in worshipping the particular god who is responsible for their desired purposes. In ancient Greece, Ares was the god of war, Hera was the goddess of childbirth, Hermes for the travellers etc. In India also Godess Lakhmi was for wealth, Godess Swaraswati for knowledge, Kartikeya was God of war etc[49].

Polytheistic Gods had physical features like human or animal and there were stories associated with each of the Gods and obviously those stories and their features were supernatural and far more superior compared to humans so that Gods could be dependable as rescuers.

Polytheism prevailing in ancient Greece and ancient Egypt do not exist to day. Yet Hinduism a polytheistic religion, is still widely followed mainly in India and South Asia[49].

There was another religion existing in ancient time which was called Pantheism that propagated the ideas "Everything in universe is divine and nothing but God." When Polytheism was prevalent in west and South Asia, Pantheism gathered momentum in Asia. Buddhism, Taoism, Confucianism were all different forms of pantheistic religions. The principal preaching was that human can find inner peace through practices like meditation and discipline and then he can integrate himself with the peace of the world[49].

With time everything changes and human minds could enlighten themselves with newer ideas and philosophies, new beliefs new religions. Such a new religion emerged called Monotheism in places where Polytheism was dominant, i.e. in Europe(Greece for example) and in Middle East(Egypt for example)[49].

In Polytheism, there were large number of deities having relative level of importance. At one time, question came in the minds of people, "who is the supreme?" They tried to provide the answer by fixing the supremacy on some of the personified deities. For example, 'Zeus' in ancient Greek religion was God of Gods, 'Brahma' was the creator of the universe and supreme God in Hindu religion, 'El' was the supreme God in Canaanite society in ancient Middle East[49].

Monotheism is the belief of one supreme God. One of the earliest monotheistic religion was Zoroastrianism founded by Zarathustra in ancient Persia in 600 BC. Zarathustra was pioneer in converting people from Polytheism to Monotheism. In Zoroastrianism, Ahura Mazda was the supreme God and believed to be the creator of the universe and all the creatures on earth[49].

Abrahamic religion, a prime Monotheistic religion started in 2000 BC in Middle East. Judaism was one of such religions whose prophet was Abraham. There were other religions, like Christianity, the messenger of which was Jesus and came about in 33 CE. Islam was introduced by prophet Muhammad in 622 AD. The commonality in these three monotheistic religions is that each one propagated stories involving ordinary human being called on by one supreme God, 'Yahweh' in Judaism, 'God' in Christianity, 'Allah' in Islam to convey message of the supreme to the common people of the world. His message was that there was one supreme God who had created everything in universe and only God was to be worshipped. Here one thing is important to note that unlike Polytheistic Gods, the supreme God had no human features and is omnipotent, omnipresent, omnibenevolent and omniscient[49].

It has been observed that though the goals of spiritual practice of each religion are not identical, they have lot of things common. All religions are connected to same ultimate reality and lead the followers to common goal. By God or Ultimate Reality or an ultimate principle we mean that it can be one and only one, whatever may be the beliefs people hold about it — be it described unique or multiple forms, personal or impersonal, absolute void or absolute being, and irrespective of the name by which it is called.

One of the most common observations is that the people attain the highest human qualities who have reached their goals may it be enlightenment, self-realization, liberation from earthly desires or salvation. Regardless of their religious beliefs, such people can successfully impress others by their personally acquired elevated state of mind.

It is often claimed that world religions form a unity, which provides the right perception on ultimate truth. Another suggestion is to consider different world religions as alternative paths to the same final destination, the ultimate reality. Yet there are exceptions to this conception also. Christianity and Islam each claims that it is showing the only right path to God and thereby try to claim that world religions are not pieces of the same puzzle or alternative paths to the same goal.

Swami Vivekananda in his speech at London nicely elaborated about this aspect. He said, "Everyone must be judged according to his own ideal, and not by that of anyone else. In our dealing with our fellow beings we constantly labour under this mistake, and I am of the opinion that the vast majority of our quarrels with one another arise simply from the one cause that we are always trying to judge others' Gods by our own, others' ideals by our ideals, and other motives by our motives." He also stated that "In judging of those ancient religions we must not take the stand point to which we incline, but must put ourselves into the position of thought and life of those earlier times."

He had reminded for restraining the fanatic religious approach by saying, "We must not forget that there will come men after us who will laugh at our ideas of religion and God in the same way that we laugh at those of the ancients."

Buddhism and God

We use the term 'God' to designate a supreme power who is considered to be the creator of the universe, the chief law giver for the humans, concerned with the welfare of His creations and grants 'Mokhsha' or salvation for those who follow His dictates. Different religions have different conceptions about Him and Buddhism also has different perception about God.

The idea of personal deity or a creator of universe of eternal and omnipotent nature is incompatible with Buddha's teachings[50].

Though there is no explicit metaphysics in Buddha's teaching, there is good deal of it in implicit form[10]. Though there is no metaphysical aim in his teaching, a metaphysical view is quite visible in the doctrine of Buddhism. Sometimes Buddhism is described as an atheistic religion, either in a rational sense by the thinkers or in a disparaging sense by people of theistic belief. In one way Buddhism can be described as atheistic, as it denies the existence of an eternal, omnipotent God or godhead who is believed to have created and is sustaining the world. The word "atheism," however, like the word "godless," conveys a number of disapproving connotations or implications, which in no way is applicable to Buddhism[50].

After becoming enlightened, Buddha preached about the ways to eliminate suffering by understanding the true nature of the world. To him, the knowledge was important so long it remained practical. He rejected the ideas related to God, the nature of the universe, and the afterlife and inspired his followers to focus instead on the basic truths by which they can free themselves from suffering.

Though it was separated from its original religion Hinduism, it has lot of things common with humanism and atheism, establishing it as a philosophy dealing with ethics to be followed in human life. Buddha did not consider the existence of God as important. Rather he claimed to emphasize on right way to escape suffering of life and attain enlightenment.

This enlightenment is the attainment of the state of nirvana. This word means coming out of all earthly experiences, to surpass all obstruction to mental freedom or to blow out of existence, the ultimate cessation of suffering and the final eradication of greed, hatred and delusion. Entering nirvana means achieving the ultimate state where one's all egos are extinguished. Their life merges in the vast sea of awakening like a drop of water. Buddha's original teaching was that nirvana was not the union with God. His system has no place for deity or personal self, rather is a state of being. The Absolute is completely impersonal, and salvation is attained solely by own effort.

To enter into state of nirvana, conception of reincarnation is introduced as the process to give one enough time to a Buddhist to develop his qualities and practices. The Buddhists desire to enter into the state of Nirvana, but there is no clear arguments or teaching that disclose what occurs beyond the death. Even Buddha himself was not convinced about life beyond death. He did not preached about the afterlife and the philosophical speculations still can be argued today. Buddhism offers neither assurance of forgiveness or eternal life. The facts of life after death still are an unresolved mystery in Buddhism.

Buddha taught, "I had no concept of self, or of a being, or of a soul, or of a person, nor had I any concept or non-concept."

The Buddhists do not believe in God because there is no concrete evidence to prove the idea of God in spite of the research on God for thousands of years. Moreover Buddhists are of the opinion that even belief in God cannot provide guarantee of peaceful, meaningful and happy life.

Theists have a general belief that God acts as a bridge between the individual and the ultimate reality. Whereas, the Buddhists completely refute this idea and contemplate that nobody else except the Karmas of the individual that drive his destination. Even Buddha cannot implore or pardon by interfering in the Karmic process[51]. According to Buddhist ideology, any one can achieve enlightenment or become Buddha and escape from the illusions of the surrounding world following the process of mental discipline and the methodology to attain nirvana.

Islam and God

According to the Qur'an, the life containing universe is solely created by the Almighty God—Allah—and He is the Supreme Ruler of the universe. A few verses on this score are:

"God has created everything. He is the custodian of everything. The keys of the heavens and the earth are held by the God."[52]

"No creature is there scrambling on the earth, but its facility rests on God. He knows where to lodge it or where is its source." [52]

"God is creator of Heaven and Earth, and whenever He desires something, He says to it: 'Be' and it is."

There are many other *ayahs* (verses) which attribute that all that is on the Earth and in the Heaven is created by Almighty God.

The scholars are of the opinion that probably the Arabs' knowledge of God (Allah) was transmitted through word of mouth from the time of Hagar and Ishmael to Muhammad (2000 BC-610 AD). Qur'an is believed to have become the final revelation rescinding all former ones[53].

Even during pre-Islamic period, the name 'Allah' (from 'al-Ilah'- the god or 'al-Liah'-the one worshipped) was very much used. It was rather a title than a name and, was used for different form of God. While the rituals are still performed addressing the respective deities, Allah is seen as the creator of the world, the superior Lord, the father of all and impartial, impersonal. Allah has become the Islamic substitute for any idol[53].

It is presumed that parts of the Bible were translated into Arabic before or at the time of Muhammad and the name Elohim (Greek or Hebrew terms for God) etc. was represented as Allah. Therefore it has become quite logical to the researchers to draw inference that from the pre-Islamic concept under the monotheistic influence of the Jews and Christians, the Qur'anic Allah emerged, and the multiple deities were replaced by the one[53].

Allah, His unanimity, His Absolute Power, having all essential attributes of an Eternal and Almighty Being, is the vital part of the Muslim religion, and is expressed as Nafiwa-Isbat means 'There is no deity' known as the Nafi, 'but Allah', as the Isbat, or that which is established, the terms Nafiwa-Isbat being applied to the first two clauses of the Muslim's Kalimah, or doctrine[55].

To the Muslims the most prominent attribute or feature of Allah, is Allahhuakbar-Allah is greater! Allah is power! He is the sovereign that is above all He is able to do as he pleases. He is limitless[55].

The fundamental base of faith in Islam is to believe that "La ilaha ill Allah" (in Arabic)

["there is no idol worthy of worship except the One Absolute and almighty God"]

As per Qur'an, Allah is forgiving and kind. He is Loving, and sensible. He is the Creator, the Healer and sustains the world. He Guides, Protects and Forgives[54].

God and Christianity

Christians believe in the Trinity i.e. God as Father, Son and Holy Spirit[56]. This should not be misunderstood as belief in three separate Gods. Christians believe that God took human form as Jesus Christ and that God is present today through the work of the Holy Spirit and evident in the actions of believers. Jesus described God as Father. In prayers He used the Aramaic word *abba* ("father") for God, which is otherwise not common in religious discourse in Judaism; it was usually used by children for their earthly father. This father-son relationship became ideal for the relationship of Christians to God.

Faith in the Son, in this case Jesus, also brought about a oneness with the Father, the God. The Son became the mediator of the glory of the Father and to convey the same to those who believe in him. In his prayer Jesus said: "The glory which thou hast given me I have given to them, that they may be one even as we are one, I in them and thou in me, that they may become perfectly one." In the prayer to his Lord, Jesus taught his disciples to address God as "our Father."

The Christian conception is that God is the creator who also sustains, governs and cares for the world. "God is light" means that in Him there is no darkness at all i.e. no darkness of error, no darkness of sin, no darkness of moral imperfection, no darkness of ignorance or of intellectual imperfection of any kind. "God is omnipotent" means that God can do all things, nothing is impossible for Him. As per Bible, God said, "Let there be light", and there was light.

The entire Bible is full of description of the infinite and esteem holiness of God. 'God is holy' is considered to be the fundamental truth of the Bible, the Old Testament and the New Testament and of both the Jewish religion and the Christian religion. The absolute, unqualified, and uncompromising holiness of God is the prime Christian conception of God.

There is another thought in the Christian conception of God and that is His eternity. God is eternal[57]. His existence had no beginning and will have no ending; He was there, He is there and He shall be there. God is always present in space and time. In Isaiah, "God is the Lord, the everlasting Lord, the Creator of the ends of the earth. He will not grow tired or weary, and his understanding no one can fathom."

One more Christian conception about God is: 'There is but one God'. The unity of God comes out again and again in both the Old Testament and the New Testament[57].

God and Jainism

'Jina' with 'ji' meaning 'to conquer' has been considered to be the origin of the word Jainism which pronounces the processes by which one can successfully subdue his passion and obtain mastery over himself[10].

Though independent of Buddhism, Jainism resembles it in so many aspects-its denial of the authority of Veda, its pessimistic outlook on life and its refusal to believe the supremacy of god. Jainism recognizes the permanent entities like the self and matter as it is observed in Brahmanism. In a word, it is a theological mean between Brahmanism and Buddhism[10].

Unlike Buddhism, the spread of Jainism is restricted in India. Strangely, it is seen wider outside the area of its birth especially in the west and south, than within it.

The Jainism has categorized the world in two classes of things-Jiva and Ajiva or conscious and unconscious or spirit and non-spirit. Jiva resembles atman or Purusha of other school of thoughts originated in India. Jiva specifies the living or animate things. The number of Jivas is infinite, all being alike and eternal. As per Jainism, the Jivas not only exist, the act and are acted upon. The jiva is characterized by infinite intelligence, infinite peace, infinite faith and infinite power. But when it unites with matter which composes Samsara, these features of Jiva are shadowed, though not destroyed. One of the very peculiar features of Jainism is the belief in the variable size of Jiva in its observed condition[10]. Jiva is capable of expansion and contraction according to the dimension of the physical body with it is associated at certain moment of time. The non-spatial character of Jiva is affected by the association with matter. This unlike the thoughts of the other religions prevailing in other Indian philosophies[10].

The individual Jiva becomes omniscient and aware of all objects accurately when the process of enlightenment culminates and all the obstacles are broken down. Since such state is arrived by itself without any external help it is called soul-knowledge and the state is 'kevala-jnana'. Mahavira is believed to have attained such state at the end of long period of his atonement[10].

Ajiva is comprised of space, time and matter (i.e. akasaa, dharma, adharma, kala and pudgala). They are different from Jiva because they lack consciousness and life. The kala is infinite but have cycles having two eras of equal duration. One is descending era when virtue gradually decreases and the other is ascending era to follow a reverse trend[10].

One of the characteristics of Jainism is its practical teaching of which the main feature is discipline. It insists not only on enlightenment alone or on conduct alone but on both. Of the various virtues to be inculcated by the Jains, 'ahimsa' occupies the foremost place. Jainism has no belief in god though it does in godhead. To Jains, every liberated soul is divine and there can be any number of such liberated souls as there is no deduction in the number. If by god we mean some supreme personality who has created the world,

Jainism must be atheistic. As per Jainism, If the god needs to create the world, it means he feels a desire which is inconsistent with his needed perfection as the supreme. So as per Jainism, there is no god and the world was never created. The main aim of the Jainism is the perfection of the soul and not the interpretation of the world[10].

God and Hinduism

A large number of deities have got place in Hindu mythology with 'Ramayana' and 'Mahabharata' the two of the most-read Hindu epics. The 'Bhagavad Gita' is the holy text of the Hindus, which is believed to have all the answers to the questions posed by the universe.

Hinduism is not simply a religion, rather it is a culture and shows its followers a way of life. To attain the perfect balance between the mental and the physical being, Hinduism preached a process of discipline called 'Yoga' to be followed by individual to attain a state of mind for further realization of the divine truth. The Trinity,'Brahma', 'Vishnu' and 'Maheswar' are the most adored gods of this culture. 'Brahma' symbolizes creation, 'Vishnu' sustains this creation and 'Maheswar' is the destroyer of the evil.

The world was formed as the divine trinity transcended into 'Avatars' or forms of being. Vishnu as described in Hindu mythology, is believed to have ten forms or Avatars in 'Garuda Purana' and another ten forms in nu's instructions, which he delivered to his carrier 'Garuda', who was the king of birds.

Unlike the concept of linear time as observed in some other religions, Hindu concept of cyclic time, such as Yugas or eras, considering the world running through gradual stages, has helped to avoid the conflict between creation and evolution of the Universe. As per the prediction there are four stages that the universe has to go through, highlighted as the Kaala Chakra, the wheel of time. The first 'Yuga' was the 'Satya Yuga', wherein the whole world was governed by the Gods, in absolute truth. The world was then an ideal manifestation of mankind. The forms or avatars of Vishnu in this era was 'Matsya', the fish, 'Kurma', the tortoise, then 'Varaha' the boar and the last one of this era was the 'Narasimha' who was half man and half lion.

Then came 'Treta Yuga' that symbolizes the morality. The Vishnu's avatars in this era are 'Vamana', the dwarf, 'Parashurama' with an axe and 'Rama' the King of Ayodhya. The greatest epic 'Ramayana' was conceived in this era. 'Dwapara Yuga' was the next era when Bhagvan Vishnu took the avatars of 'Balarama' along with his brother 'Krishna'. The great epic 'Mahabharata' was story of the events during this era. It is believed that this Yuga saw its end when Lord Krishna returned to his eternal home at 'Vaikuntha'. The fourth 'Yuga' known as 'Kali Yuga' which is as per Hindu mythology is the age of vice. According to Hindus the world is presently in the 'Kali Yuga'. The death of Krishna marked the beginning of this age. The believers of Hinduism think in this stage of transformation of the universe, the world is degenerated spiritually as people are moving away from the God in their mental character.

The Hindu mythology in the form of 'Puranas' and 'Vedas' has been written in Sanskrit. The epics of 'Ramayana' and 'Mahabharata' have also been scripted in Sanskrit. 'Shrimad Bhagwad Gita'is the compilation of the conversation between 'Lord Krishna' and 'Arjuna' on the battlefield of 'Kurukshetra' has become the most sacred document to the Hindus.

All the God's creations including nature are manifestations of Him. He is all pervading within and without his creations, and also observing it externally. Hence all animals and humans in the universe have a divine element in them that is covered by the ignorance and illusions of material or irreverent existence.

As happened to the Polytheist religion, in Hindu religion also there were a number of Gods and with stages of spiritual evolution and metaphysical thought the names of various gods and goddesses came into the Hindu scriptures like Veda: Mitra, the Sun; Varuna, the god of night and of the blue sky; Dyu, the Sky; Prithivi, the Earth; Agni or fire god; Savitri, the brilliance; Indra, the master of the universe; Vishnu, the bearer of the three worlds; and Aditi, the mother of all other gods[58].

The belief in the plurality of gods was a distinctive feature of early Vedic religion. Gradually, with time, this view is faded losing its attraction. Vedic Indian, dissatisfied with the old mythology and not being convinced to relate the existence of gods merely for the causes of natural phenomena, attempted to find out the ultimate cause[10]. To reduce many gods of early mythology to one, they had the easiest way to elevate the most imposing of them to the supreme position [10]who would be seen as controller all other divine entities. It caused the progress of man's concept of God or the ultimate Reality from polytheism to monotheism. We find the same concept in the Rig Veda, where the sage asks:

'Kasmai devaya havisha vidhema?

[To what god shall we offer our oblations?][59]

'Ko dadarsha prathamam jayamanam?
mastahnvant yaadnastha bibhurti
Bhoomya asurasrugatma kva switko
vidwaansamoopa gaatprashtumetat

[To achieve the spiritual knowledge of the self is the main goal of man's life. What is the main reason for the manifestation of matter? Man must obtain the knowledge of the body, the strangeness of the blood, muscles, bones, etc. within it and their being different from the soul (or the spirit)][60]

The idea of one God has been nicely elaborated in first mandala of Rig Veda:

Índram mitráṃ váruṇ amagnimāhu/
Rátho divyáḥ sá suparṇó garútmān/
ékaṃ sádviprā bahudhâ vadantya /
agníṃ yamám mātaríśvānamāhuḥ//

["They (the men of wisdom) call him Indra, Mitra, Varuna, Agni, and he is the heavenly, noble-winged Garutman. The Reality is one, but sages call it by many names; they call it Agni, Yama, Matarishvan (and so on)."][58]

There may be numerous and different names but all denote the one God which has also been expressed in Viswakarma Sukta:

yo devaanam namadha eka eva
tam samprasnam bhuvana yanty anya.

The best conceptions of God in the whole of Vedic literature is available in the last chapter of the *Shukla Yajur Veda Samhita,* which is also known as the *Ishavasya Upanishad.* It is said :

isavasyam idam sarvam yat kinca jagatyam jagat
tena tyaktena bhunjitha ma gridhah kasya svid dhanam

[Everything animate or inanimate that is within the universe is controlled and owned by the Lord. One should therefore accept only those things necessary for himself, which are set aside as his quota, and one should not accept other things, knowing well to whom they belong.][61]

God created this world, then entered into everything. In the Taittiriya Upanishad the idea is depicted more convincingly:

Idam sarvamsrijat| yadidam kincha|
Tatsrishtaya tadevanuprabishat| tadanuprabishva|
Sachha twachcha bhabat ||

['It created all this that exists. Having created (that) It entered therein. Having entered, It became the formed and the formless.'][58]

CHAPTER V

QUEST FOR ELEMENTARY PARTICLES OF MATTER

The universe, the conglomeration of the galaxies has remained a subject of mystery generating enormous inquisitiveness about its origin in the human minds over the centuries. It is not only to know how the universe was created but also to explore the magical phenomena from the core of which the matter was created out of nothing from the state of singularity.

Even before 100 years from now, the universe appeared very simple to mankind. The people were thinking that it was eternal, unchanging, a single galaxy containing only a few million visible stars. But with the advent of new powerful instruments like Hubble Space Telescope to observe the universe and new scientific ideas of the 20th century, starting from Einstein's general relativity to modern theories of the elementary particles, the picture today is totally changed and lot of information about the origin and status of the observable universe is available. As mentioned earlier, the scientists are now of the opinion that cosmos originated about 13.7 billion years ago with the big explosion. After a fraction of a second of the beginning, the universe was a hot, soup of plasma of the most elementary particles, quarks and leptons. Thereafter it was gradually expanded and cooled, structure developed layer by layer and other particles were created. Subsequently gradual formation starting from nuclei then atoms, clustering to form stars, galaxies, constellations of galaxies, and finally super clusters were formed. It is now said that the observable part of the universe is inhabited by 100 billion galaxies, each containing 100 billion stars and perhaps a similar number of planets. The first microsecond after big bang was the formative period for the universe when matter came to dominate over antimatter, the seeds were planted for galaxies and other structures, and dark matter was supposed to have been created which was some unidentified and mysterious material that held the universal structures together. The Scientists are of the view that the universe continues to expand at an accelerating pace, driven by dark energy, whose gravitational force repels and not attracts.

When the creation of universe has remained as exploration of the macro world accepting the domination of theory of relativity, there has been another stream of study in the micro world evolving theories like quantum physics to harness interior of the nucleus of the atom constituting the matter and to find the origin.

Ancient physics

In spite of frequent fights between the city-states of Athens, Sparta, Thebes and others, the Greek civilization climbed to a crest which allowed the Greeks to excel in arts and philosophy that gave the birth of a few of the greatest philosophers of all time[62].

The ancient physics started its journey with the zeal to find out the mysteries behind the creation of the universe and the laws that governed the universe. The mathematicians like Pythagoras, Euclid tried to bind those laws in mathematical equations that made further journey to the world of physics easier.

Thales of Miletus was the first physicist who believed that world, though are gifted with many materials, is built of only one element, the water, called 'Physis' in ancient Greek. He was also aware of the attractive power of the magnets and rubbed amber. The first explanation of the natural phenomena, perhaps could be brought out first through his postulations and the natural activities could come out of the divine domain into realm of natural laws and explanations[62].

Anaximander, another Greek philosopher, disagreed with Thales and proposed that a substance called 'apeiron' as the building block of all matters and not water. Empedocles proposed that the building blocks were four instead of one, i.e. earth, water, air and fire. Around 500 BCE, Heracletus gave the idea that principal of change was the only basic law controlling the universe and nothing could remain in the same state indefinitely, thereby 'time' became an important aspect of the natural laws[62].

Democritus, a student of Leucippus (around 500 BCE) announced the first atomic theory that matter cannot be divided indefinitely and a time would arrive so that further division would not be possible. He called this indivisible particle, the Atom (a-tom—not cut)[62].

Around 350 BCE, Aristotle tried to consolidate the knowledge of physics so far gathered and proposed a set of theories that provided the fundamental basis science for next thousand years[63].

Aristotle tried to explain events like motion and gravity with his theory of elements which was a new conception to ancient physics that spread further to medicine and alchemy. Aristotle suggested that matter was made of five elements, earth, air, water, fire and one invisible element, he called it as 'aether'. Aristotle's ideas continued to govern scientific thoughts until the greats like Galileo and Newton entered into the scenario[62].

The ideas given by the Greeks were reigning the world of physics and developed the base of modern principle of physics. In fact, very few new developments took place in the centuries following the Greek period till the tremendous intellectual growth during Renaissance acted as catalyst to drive the acquired knowledge of physics to get further enriched and started challenging the previous doctrines of the scientific authorities of the ancient Greek philosophers. Copernicus' theories were such revolutionary in nature that it ended the old era of scientific understandings and created an overall surge in new thoughts in the scientific and philosophical world beginning around 1500 AD[63].

Thereafter, Galileo entered into the scene and wanted to replace old assumptions by new scientifically established theories. He proposed the celestial theories and his works on mechanics helped further formulations of laws of motion. Tycho Brahe provided accurate celestial data which helped Kepler in developing theory of elliptical planetary motion [63].

Physics in nineteenth century

Nineteenth century witnessed brilliant works of the scientists one after another that started revealing the structure of matter in more comprehensive ways. The theory of George Stoney in 1874 gave the idea of **electron** and estimated its mass. In 1895[63], a mysterious and highly penetrating radiation was discovered by Roentgen which was called X-ray.

One year after i.e. in 1896, Thompson discovered, using the cathode ray tube, that the negatively charged particle was electron which was considered as the first fundamental particle. It was clear by this experiment that atom is divisible and not a fundamental particle. Thompson, thereafter, proposed that an atom was a sphere with homogeneously distributed positive charge in which electrons were located as the raisins in a pudding cake[63].

In 1896, French physicist, Becquerel noticed that X-ray was causing certain material to fluoresce. He then tried to test whether the phosphorescent materials emitted X-ray when they were glowing. Putting some of these glowing materials, after getting them excited exposing to sun light, on top of covered photographic plates, with utter confusion he found mixed result i.e. in some cases the plates were affected and for others it was not. But when he put the materials giving positive result on uncovered photographic plates on a cloudy day without giving chance to those materials to get excited through sunlight, he was astonished to watch that the plates were blurred. He thus discovered that those materials were uranium compound radiating such rays which he called the 'uranic rays'[63].

Rutherford further studied this uranic rays and found that they were having two components, one was positively charged, very heavy and easily absorbed in matter. The other was negatively charged, much lighter much penetrating and not so easily absorbed in matter. Rutherford called these two components as **'a'** and **'b'** particles.

In 1898, Pierre and Marie Curie, the French physicist couple later termed these uranic rays as radioactivity. Becquerel, by his continued study with uranic rays detected that the particles acted like electron of very high energy. Further study with uranium revealed that during radiation, a third type of particle was also emerging which was more penetrating than the a or b particles and called it **'g'** particle[63].

Concept of nucleus of atom

While Rutherford was experimenting with a-particle beamed against thin gold sheets, he found that some of the particles were bouncing back as if they had been obstructed by a massive, positively charged particles located in a small region of the atom. This observation lead the discovery of atomic nucleus in 1911 and an atomic model was proposed which was like a miniature solar system where the nucleus was representing the Sun and the electrons as the planets[63]. The attraction between the nucleus and the electrons was the electric coulomb force by which the particles are bound within the atom. The approximate sizes of atom and its constituents are given at Fig. 5.1[7].

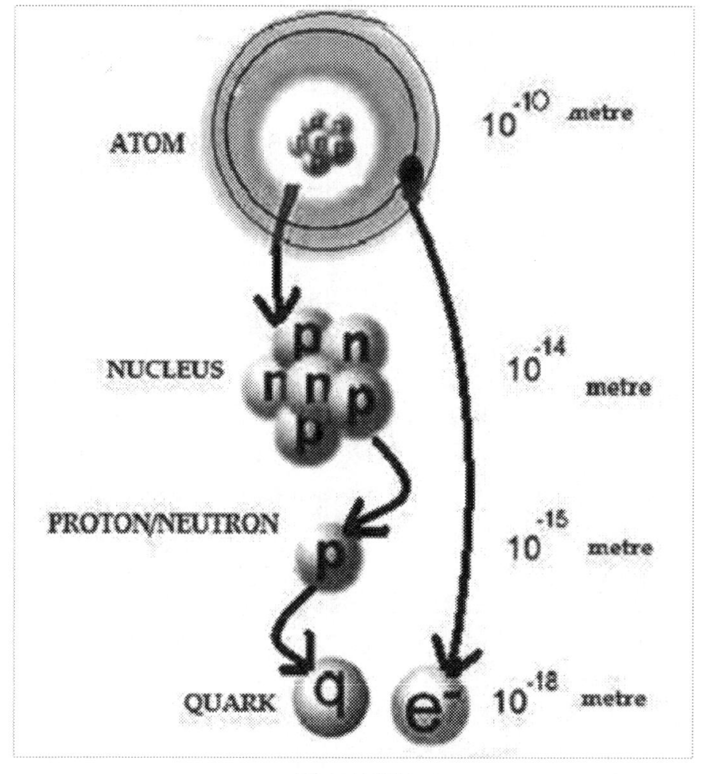

Fig. 5.1[7]

Even accepting this atomic model, the physicists observed that something was yet to be explained. As per classical physics, the accelerated negative charge while moving around the nucleus, should have emitted electromagnetic radiation and thereby loose energy gradually. Thus at one time, the electrons would fall on the nucleus and hence the atom would be unstable. Also as per classical theory the frequency of radiation so emitted by the electrons would change continuously. But the observation of spectrum analysis did not support this phenomena[64].

Twentieth century and modern physics

At the beginning of twentieth century, there was an atmosphere of some sort of complacency that most of the fundamental principles of nature were known i.e. atoms are the solid building blocks of nature, Newtonian laws of motion were sufficient to describe the moving objects and many other queries related to events happening in nature could be explained with the available theories and postulations[62].

Introduction to quantum theory

But the belief of the people got abruptly shaken when Max Planck suggested the quantum theory and Einstein's startling theory of relativity replaced Newtonian mechanics. The scientists started thinking that there were lot more to be known and their knowledge about nature was not complete. The fundamental concept of physics got changed and a new field, the quantum physics took over the reign of major scientific adventure thereafter.

In 1900 Max Planck discovered that the radiation spectrum of black bodies occurs only with discrete energies separated by the value hv where v is the frequency and h is a new constant, called Planck constant. The fact that the absorption and the emission of energy is not continuous is in conflict with the principles of classical physics. After a few years, Einstein used this discovery in his explanation of the photoelectric effect. He proposed that light waves were quantized, and that the amount of energy of each quantum of light was exactly hv.

Bohr's atomic model

It was Niels Bohr, the Danish physicist who came forward to solve the problem being faced by the physicists at that time and proposed a new atomic model in 1913 for hydrogen atom using quantum theory. The proposal says that the electrons are moving around the nucleus in stationary orbits and the electrons in the outer orbit determine chemical properties of the atom. The electron in any such orbit remains there unless excited. Bohr used data from the hydrogen spectrum. The line spectrum of hydrogen indicated that only certain energies were being allowed for the electrons in the hydrogen atom.

The changes in energy levels of electron in hydrogen, emits certain wavelengths of lights. Therefore, the energy levels of the electrons in the atom is quantized. On excitation, say by heating, it moves to higher orbit and when comes back to its original orbit, energy is emitted in form of light of frequency v and the energy emitted is hv where h is Planck's constant and hv is the energy difference of the stationery orbits between which the electron is transiting. With this simple proposition, Bohr could explain the spectrum of hydrogen atom[64].

Rutherford continued his work of bombarding things with a-particle and in this process he produced a hydrogen nucleus which was called **Proton** [63].

Wave-particle dualism

In 1905 Einstein suggested that light, can be thought of as a particle also in addition to its well known characteristics of an electromagnetic wave, which now we call the photon. In 1923, the French scientist Louis de Broglie proposed that electrons having particle-like nature, could also be thought of as some sort of wave. de Broglie was doing his PhD thesis at that time. His supervising committee didn't know what to make of this strange proposal and asked Austrian physicist Schrödinger, who stated that the idea was "rubbish!" The committee went to Einstein, who suggested that they should give Broglie, his PhD, as "there might be something to it." So de Broglie obtained his PhD, and in 1926 Davisson and Germer actually saw electrons demonstrating an interference pattern. Louis de Broglie proposed his wave particle dualism theory in 1924 stating that any particle moving with momentum p, will be associated with a wave of wavelength $\lambda = h/p$[65].

The conception of the wave particle dualism got further support from the German physicist Heisenberg who postulated uncertainty principle in 1924. As per this principle, it is not possible to define simultaneously the position and the momentum of a particle with absolute precision. The uncertainty principle also says that indeterminacy of momentum multiplied by the uncertainty of position is of the order of h, the Planck constant. In macroscopic world, where we are dealing with bigger objects, h becomes negligibly small compared to typical values of the system, we go back to classical orbit, classical concept of a particle and uncertainty principle loose its relevance. But as we go for much smaller objects when magnitude of h becomes comparable to the relevant dimension of the system, as it happens in case of electron having a very small mass of 9 x 10-31 Kg, the uncertainty principle comes into the picture and classical idea vanishes.

So for electron's case, if it is De Broglie's wave and not a particle as per classical concept, classical trajectories do not exist. Then how it fits with the old concept of orbit? It is Max Born, who brought the idea of wave function associated with the particle which would give us the probability of finding the particle at some particular location and said that in the region where amplitude of the wave is not zero, there is possibility of locating the particle but not with absolute certainty. Where the wave amplitude is zero, it is impossible to locate the particle. Thus the quantum mechanics which replaced the classical mechanics is actually a probabilistic theory.

Schrodinger's atomic model and quantum numbers

Schrodinger applied the idea for electron in the stationery orbit around the nucleus in an atom. These stationery orbit should correspond to such a perimeter which accommodates integer number of electron's wavelength λ. This is possible only for certain values of λ and hence certain value of the momentum or energy. The discrete energy values appearing in the atomic spectrum could be explained accordingly.

Schrodinger adopted the ideas proposed by de Broglie, the theory of uncertainty principle of Heisenberg and ideas developed by others and combined those together in a single equation that is called Schrodinger's equation[69]. In 1926 the wave equation proposed by Schrodinger accurately described the dynamics of the micro world of atoms and molecules of the matter. Thereafter it has been universally accepted as one of the greatest achievements of the 20th century physics.

Bohr's atomic model was one dimensional in the sense that the orbits around the nucleus was characterised by one quantum number 'n' depending on the size of the orbit, where n = 1, 2, 3, . . . and later on n was treated as principal quantum number.

Bohr's theory was not able to explain all the spectral lines of the hydrogen like atoms. It was observed that the spectral lines were not homogeneous. Modifications of the theory, therefore was necessary to explain all those lines. Arnold Sommerfeld, the German Scientist proposed that the electron orbits need not be spherical, rather the electrons are moving in specific elliptic orbits. For this he introduced conception of a secondary quantum number called azimuthal quantum number (*l*) which defined the shape of the orbits. For any principal quantum number n, there are n kinds of shapes but characterized by different energies.

Another improvement was made to Bohr's model by the observation that the orbits need not be in the same plane. These planes can be oriented in space in some defined directions. The conception of another quantum number was introduced to define the orientation of such planes and this was called the magnetic quantum number(m). The electron orbiting in its orbit generates magnetic field and when the orbiting electron is placed in an outer magnetic field, the orbit of the electron places itself such a way that the direction and sense of the magnetic field caused by the moving electron follows that of the outer magnetic field. To deflect the orbit system to other position some more energy needs to be given to the system. Sommerfeld showed that there would be some defined number of possible orbit's position and each position would have a bit different energies. He calculated that such number of positions would be 2*l+1. So for *l*=0, number of positions will be only 1 i.e. m=0. For *l*=1, m will have three values, +1,0,-1. It was observed that the orbits of different energies for the same value of n explained the splitting of lines-the Zeeman effect.

Schrodinger's model prescribed for three dimensional space for the electron. At this juncture, we should clear the conception about orbit and orbital which sounds to have similar meaning. But actually it is not. The planet moves around the sun and the trajectory path can be plotted accurately and the path is called the orbit. In the planetary model of atom also, had electron been on a orbit around the nucleus, the position of electron could have been located exactly at different time on the trajectory of the orbit. Now drawing reference of Heisenberg's uncertainty principle, it can be said that though we can have knowledge of the position of the electron at some moment of time, the position at the next moment cannot be predicted with certainty. The solution of the Schrodinger's equation gives not the exact position of the electron, but the probability of finding the electron in a specific place around the nucleus. This most probable "place or location" is called an orbital. An orbital is a volume of space around the nucleus where the probability of detecting the electron will be 90%. So the electron is not orbiting but actually its presence is spread in a region of space i.e. orbital. The orbitals are also called electron clouds. The change of the orbital structure in an atom depending on value of n, l and m is explained in Appendix-C.

Discovery of neutron

After Bohr proposed his atomic model in 1913, the physicists had to wait for about eighteen years to discover that there was another sub atomic particle apart from Proton in the nucleus and that was called Neutron. In 1931, the British physicist Chadwick, while bombarding element Beryllium by alpha particle, noticed radiation of a stream of electrically neutral particles which he called **neutron**.

Further studies carried out on neutron reveals that a free neutron outside the nucleus is unstable. The half life (see Appendix C) of neutron outside nucleus is about 12 minutes only. But neutron within the nucleus is stable. However, a neutron in a nucleus will decay if a more stable nucleus results; the half-life of the decay depends on the particular isotope. If it can provide a more stable nucleus, a proton in a nucleus may capture an electron from the atom (electron capture), and get converted into a neutron and a neutrino (see later). When the neutron decays within the nucleus, the atom becomes radioactive. The beams of neutron are highly penetrating and are dangerous for living tissues.

With the discovery of neutron, the physicists defined another parameter of the atom of an element which is atomic mass number and equal to the summation of number of proton and number of neutron in the nucleus of an atom. The element Carbon has 6 protons and 6 neutron in nucleus of its atom as shown in Fig.5.2.

Dirac's equation and concept of antimatter

In 1930, in his attempt to combine relativity and quantum mechanics, the British physicist Dirac proposed his famous relativistic quantum

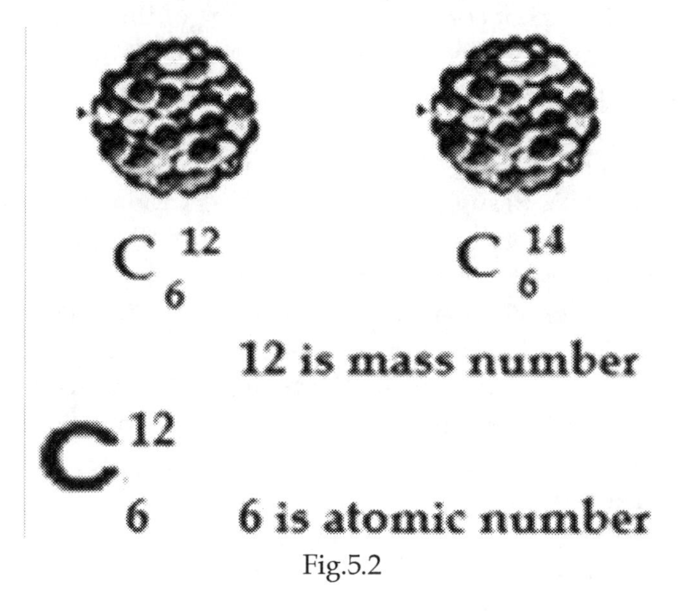

Fig.5.2

equation, most familiarly called Dirac's equation that predicted negative state of electron and proton and hence conception of antimatter came into the picture.

Discovery of neutrino

Beta particles are electrons or positrons (which is antielectrons i.e. electrons with positive electric charge). Beta decay occurs when, in a nucleus with either too many protons or too many neutrons, one of the protons or neutrons is converted into the other. This remained a puzzle to the physicists for quite some time as the electrons emitted in the process do not have same kinetic energy always and the emission is taking place with a bell-curve type distribution (see Appendix-C) of energies to indicate that the energy is not conserved in the process and the missing kinetic energy varies in a probabilistic fashion. It indicates that the nuclear energy responsible for beta decay is going somewhere other than the emitted electron.

In 1930, Pauli suggested the existence of another fundamental particle to explain continuous electron spectrum in beta decay. Maintaining his stand in favour of conservation laws he got involvement of the presence of a very light, charge less particle simultaneously emitted during the decay which was **neutrino**.

In 1934, Italian physicist Enrico Fermi utilised the conception of neutrino which Pauli suggested for the preservation of the principle of conservation of energy to explain the process of beta decay. Fermi proposed that along with electron another particle is emitted which is neutrino. This has not only strengthened Pauli's proposal of neutrino but has played a significant role later on in driving the modern physics onwards. Neutrinos are produced through radioactive interactions like that happen in fusion process in sun, supernovae, radioactive decay, and when cosmic rays collide with the Earth's atmosphere.

In beta minus decay, a neutron decays into a proton, an electron, and an antineutrino: $n \rightarrow p^{+} + e^{-} + \acute{\upsilon}$. In beta plus decay, a proton decays into a neutron, a positron, and a neutrino: $p \rightarrow n + e^{+} + \upsilon$ where e^{-} is electron, e+ is positron, $\acute{\upsilon}$ is anti neutrino and υ is neutrino. This establishes the electric charge conservation (Fig.5.3)[7].

In beta decay, binding energy is changed and it generates the energy due to mass and kinetic energy of the beta particle, the energy of the neutrino, and the kinetic energy of the daughter nucleus generated. While still obeying energy and momentum conservation, the energy of an emitted beta particle from a particular decay can be of different values as the released energy can be shared in many ways among the three particles.

Protons can be changed into neutrons or vice-versa in cases of proton decay or neutron decay or electron capture. In each decay the atomic number gets changed so that the parent and newly formed atoms are different elements. In all three processes, the mass

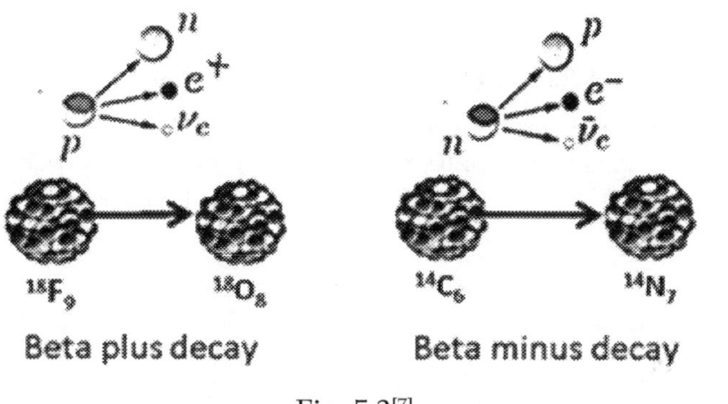

Fig. 5.3[7]

number of nucleons remains the same, while both proton number (atomic number) Z, and neutron number *N* is changed (increased or decreased by 1).

Discovery of existence of positron

Though the existence of the positron was first predicted by Paul Dirac in 1930 as a result of his work on the application of the special theory of relativity to quantum mechanics, the **positron** was first discovered in 1932 by Carl Anderson and this was the discovery of the first antiparticle which became a milestone to begin a new era of particle physics. Like other investigations on cosmic ray particles made by the physicists using Wilson cloud chamber, Anderson was using a cloud chamber filled up with a gas supersaturated with water vapour in a magnetic field to study the nature of cosmic ray particles. The magnetic field was to bend the trajectories. The charged particle while passing through the a thin lead plate placed across the cloud chamber, caused the vapour to condense along its path.

Anderson observed the cosmic ray tracks in the cloud chamber and compared the same with known properties of electron and proton tracks. The width of the track was found consistent with the charge of the order of electron charge and path was more bent on one side than the other. In the picture at Fig.5.4, a charged particle is seen entering from the bottom at very high energy[71].

The particle then lost some of the energy in passing through the lead plate in the middle. From the curvature of the track it could be deduced that it was a positively charged particle. Anderson then explained that the new positively charged particle carrying same amount of charge as carried by electron is negative electron and

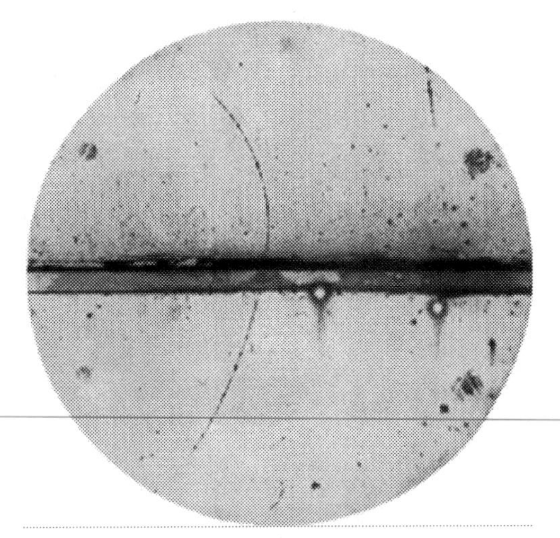

Fig.5.4[71]

named it as 'positron'. From the energy loss in the lead plate and the length of the tracks after passing though the lead plate, an upper limit of the mass of the particle can be made. In this case Anderson deduced that the mass was less than two times the mass of the electron hence the positively charged particle is not proton.

Discovery of Mesons

Yukawa, the Japanese physicist proposed in 1935 a new theory of nuclear forces in which he predicted the existence of 'mesons' as the particles that had masses between those of the electron and the proton(about 200 times of mass of electron). The necessity of introduction of a specific nuclear forces was realized soon after the discovery of the neutron, and the strong binding between the protons and neutron could not be explained as electromagnetic interactions between charged particles[76].

In order to describe the nuclear interactions between protons and neutrons which provides the force necessary to bind the nucleons in the nucleus he combined relativity and quantum theory and proposed that it was the exchange of a new type of particle interacting between neutrons and protons in the nucleus, called '**mesons**'. It was thought that the meson emitted by one nucleon would be absorbed by another nucleon and a strong force between the nucleons would be produced, analogous to the force produced by the exchange of photons between charged particles interacting through the electromagnetic force. Of course, now it is known that the strong force is mediated by the gluon (see later).

A particle having more or less same mass was discovered in cosmic ray investigation experiments in 1937 but could not establish Yukawa's prediction of meson as the particle was found having half integral spin and behaving as fermions and not boson of integral spin as proposed by Yukawa to define strong nuclear force(see Later). This cosmic ray particle was not Yukawa's particle and appeared to be a heavy electron, later on called as mu meson or '**muons**'. With further research around 1948 and thereafter three particles having masses equal to about 270 times that of electron were discovered, one with positive charge, one with negative charge and the other was neutral. These particles were called pi-meson or **pions** and found having properties identical to that predicted by Yukawa[76].

Fermions and Bosons

The physicists divide the elementary particles in two categories. One category with those particles that have odd half integer spin(1/2, 3/2 and so on) and obey exclusion principle are called '**Fermions**'. Electrons, protons, neutrons etc all obey exclusion principle and hence are fermions. In 1925, Einstein and the famous Indian scientist Satyendra Nath Bose developed a set of equations to describe the nature of the particles which do not obey Pauli's exclusion principle. Since these particles follow Bose-Einstein statistics, are called '**Bosons**' which have integer spin. There is no limit to these particles to occupy the same quantum state. All the force carrier particles are Bosons. Photon, the light particles are boson[76].

As it happens for odd and even numbers sum of odd and even numbers are either odd or even depending on how many odd numbers are added i.e. addition two odd numbers provide an even number and addition of a third odd number makes the sum an odd number. Any number of even numbers add up to an even number only. For fermions and bosons, identical things happen. When even number of fermions are added, we get boson. Addition of odd number of fermions provides fermion. Addition of any number of boson yields boson only. The nucleus of an atom may be either fermion or boson depending on the total number of protons and neutrons is odd or even.

If we see hydrogen atom, it is having two fermions, one electron and one proton but the overall atom is a boson since consists of even number of fermions. Hence one hydrogen atom can be identical to another hydrogen atom but electron in hydrogen atom cannot have same properties as another electron in the vicinity. In deuteron or heavy hydrogen atom where there are one electron, one proton and one neutron i.e. three fermions, so the deuteron atom is also a fermion.

Leptons and quarks—the fermions

The scientists have categorised the fundamental particles fermions into **Leptons** and **Quarks** each of which has antiparticle partner with same mass but opposite charge.

Leptons are the fermions which are not influenced by strong interaction force. It is further divided into two categories. Charged leptons(like electron) and neutral leptons (like neutrinos). A total of six leptons are found in three sub categories—electron, muon, tau and their respective three types of neutrinos.

Martin Perl, the American physicist, in 1975 discovered a new elementary particle known as the tau lepton. The **tau lepton** and electron are identical in all respects except that tau is much heavier i.e. more than 3,500 times than the electron but survives only less than a trillionth of a second, whereas the electron's existence is stable.

Around '70s, the physicists started thinking that protons and neutrons in the nucleus are not elementary particles as they are not sufficient to explain the large number of new particles being discovered one after another. In 1964, Gell-Mann and Zweig proposed that the so called fundamental particles protons and neutrons are rather composed of some point like objects which they called as **Quarks**. These particles, quarks, should have electrical charges equal to 1/3 or 2/3 that of an electron or proton. Accordingly, extensive experimentation started with high energy accelerators, cosmic ray investigation follow ups and other routine research processes. But the efforts were not being successful. Thereafter, a series of experiments known as inelastic electron-nucleon scattering were conducted between 1967 and 1973 at the Stanford Linear Accelerator Centre. Other theoretical and experimental advances of the 1970s later on confirmed this discovery and established a new model for the elementary particles, called standard model. Quarks are recognized today as the elementary particles like other such particles by which the matter is composed[89].

The **antiquark** is the anti-particle of a quark. these two particles are the only two fundamental particles which interact through all four fundamental forces of nature. The quarks are not observed independently but always in combination with other quarks. So it is not possible to determine the properties i.e. mass, spin and parity of a particular type of quark directly. The idea is to be obtained from the particles built up by them. The measurements indicate existence of a non-integer spin of quark (either +1/2 or -1/2), so quarks are fermions and follow the Pauli Exclusion Principle. There are six distinct type of quarks: up, charm, top down, strange and bottom. These peculiar names for quarks do not describe any aspect of the particles; they merely give physicists a way to refer to a particular type of quark. Further details on quarks is available in Appendix-C.

Baryons and mesons—the hadrons

In particle physics, a **hadron** is a composite particle made of quarks held together by the strong force (like the electromagnetic force that binds the atoms and molecules together). Hadrons are categorized into two families: **baryons** (made of three quarks) and **mesons** (made of one quark and one antiquark).

A baryon is constituted by three quarks i.e. an odd number of fermions, so it is a fermion itself. Proton consists of two up quarks and a down quark (u+u+d) as shown in Fig.5.5. and hence it is baryons.

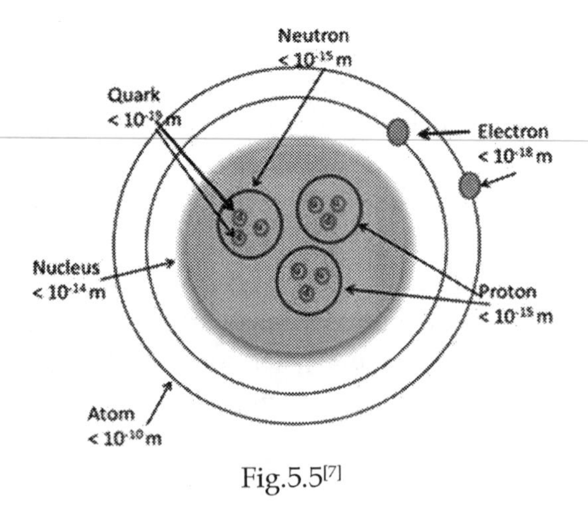

Fig.5.5[7]

Adding the charges of two up quarks and a down quark, +2/3 plus +2/3 plus -1/3, produces a net charge of +1, the charge of the proton. Protons have never been observed to decay.

The neutrons in the nucleus of an atom are also baryons. A neutron consists of one up quark and two down quarks (u+d+d). Adding these charges gives +2/3 plus -1/3 plus -1/3 making the neutron charge less and hence electrically neutral. Neutrons possess a slightly more mass than protons and an average lifetime of 930 seconds.

Mesons contain a quark and an antiquark (the antiparticle partner of the quark). Since they contain two i.e. even number of fermions, mesons are bosons. The pion was the first meson that was detected by the scientists. Pions acts as transitional particles in the nuclei of atoms, being absorbed by protons and neutrons.

The pion comes in three varieties: a positive pion (Π^+), a negative pion (Π^-), and an electrically neutral pion (Π^0). The positive pion consists of an up quark and a down antiquark. The up quark has charge +2/3 and the down antiquark has charge +1/3, so the charge on the positive pion is +1. The average lifetime of positive pions is 26 nanoseconds. The negative pion is constituted by an up antiquark and a down quark. The neutral pion consists of an up quark and up antiquark or a down quark and down antiquark so the electric charges cancel each other. It has an average lifetime of 9 femtoseconds. (femto - 10^{-15})

Standard model of particle

Except electron, most of the particles discovered subsequently like proton, neutron etc. and thought of at first instance as elementary particles, have been found to be not elementary, rather are composite particles. There has been tremendous experimental and theoretical efforts in the last century to know what is the ultimate constituents of matter and what is the process of interactions between them. Can there be any mathematical model for those elementary particles? What are the basic differences ? and theoretical efforts in the last century to know what is the ultimate constituents of matter and what is the process of interactions between them. Can there be any mathematical model for those elementary particles? What are the basic differences ? All these questions were haunting in the mind of physicists for pretty long time. The standard model of particles is the consolidation of the acquired knowledge gathered out of the expedition of getting the answers of the innumerable questions about the structure of matter[73].

Standard model and the Gravity

The standard model proposes that the materials in the universe are nothing but elementary fermions (Leptons and Quarks) which are interacting through some fundamental forces and the boson particles which are responsible for carrying such forces (see Fig.5.6A&5.6B). The four fundamental forces of the universe: the electromagnetic force, the gravitational force, strong force and the weak force. These forces are operative with different strength and extended to different areas for their actions[73].

The electromagnetic force affects any electrically charged fundamental particle (all the quarks and half of the leptons). Through this force only the lightning strikes and different poles of bar magnets attract each other. The weak force contributes to radioactive decay. It actually facilitates conversion of neutrons into protons in a nucleus. The strong force (called so because it is stronger than the weak force) is only felt by quarks. Its elastic behaviour makes the pull between two quarks stronger the more they are made apart. Each force has one or more force-carrying particles associated with it. The Standard Model clarifies the queries extremely well how the fundamental forces act on the matter particles. But the gravity, which is mostly experienced in our daily lives, could not be included in the format of the Standard Model. This has remained a big challenge to the physicists to find out a unified theory that also will include the gravity[80].

Out of the four fundamental forces, Gravity is the weakest but it has infinite range of influence. The electromagnetic force also has also infinite range but is much more stronger than gravity. The weak and strong forces are dominating only at the level of subatomic particles i.e. at micro level.

Name of the force carrier particle	Force	On which the force acts
GRAVITON	Gravity	Which has mass
PHOTON	Electromagnetic	Which has charge
GLUON	Strong nuclear	Protons and Neutrons
Z-BOSON W-BOSON	Weak force	All fundamental particles

The strong force is the strongest among all the four fundamental interactions and weak force is the weakest. See the table given above[73].

Standard model proposed that three of the fundamental forces result from the exchange of some fundamental particles 'bosons' called force carrier particles,. The particles in the matter transfer discrete amounts of energy with each other by exchanging bosons. Each fundamental force has its own boson particle responsible for the interaction between corresponding particles—the strong force is carried by the 'gluon', the electromagnetic force is carried by the 'photon', and the 'W and Z bosons' are responsible for carrying the weak force.

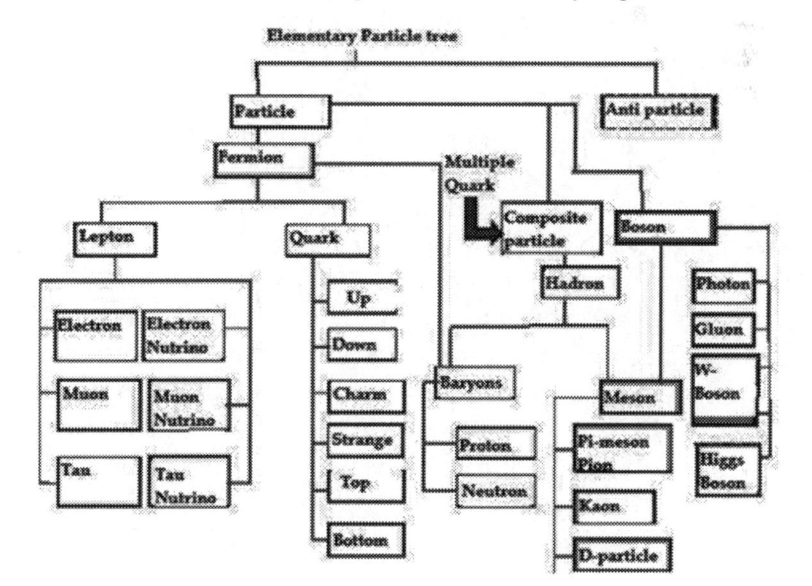

Fig.5.6A

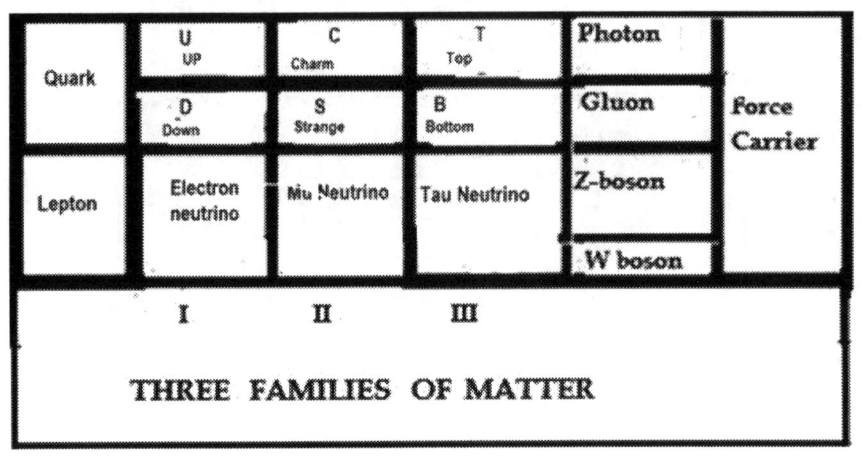

Fig.5.6 B[78]

Photons are devoid of mass and charge and have an unlimited range. Quantum electrodynamics (QED) is a mathematical model that describes the interaction of charged particles through exchange of photons.[73].

Quarks cling to other quarks because they possess a feature known as colour (or colour charge). By 'colour' we do not mean the word colour as we normally use. The name is completely arbitrary and whimsical like the names of quarks. Quarks are too much small to be visible and never have a perceptual property like colour which can be seen. The particular word was used based on the peculiar similarity of behaviour of different quarks with that of different colours of light visible through human vision. The colours of quarks in the standard model combine like the colours of normal light for example the three colours add up to white. Hence analogically all quarks have been assigned three colours: red, green, and blue and anti quarks colours are cyan, magenta and yellow. Like colours repel and unlike colours attract. Colour-anti-colour attraction is stronger than attraction of colour-colour attraction. Quark particles are bound together by gluons (fig. 5.7).

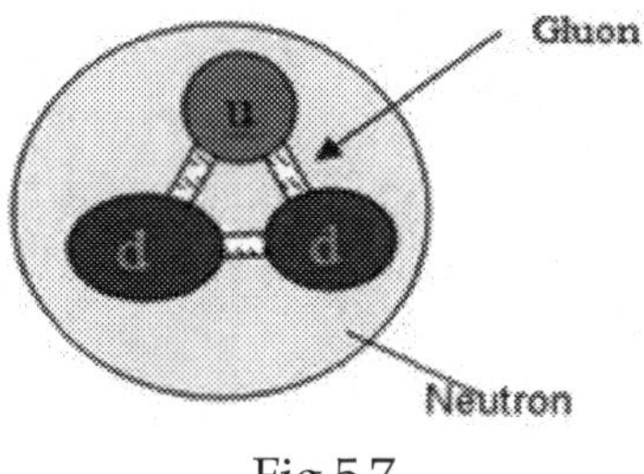

Fig.5.7

Gluons are also coloured in the same sense as quarks, but in a more complicated way than the quarks are (further details in Appendix-C).

The 'graviton' still remains as hypothetical and supposed to be the corresponding force-carrying particle of gravity (Fig27of Appendix-C).

The standard model specifies seventeen particles. Six leptons, six quarks, five bosons. W and Z bosons were discovered in 1983, the top quark in 1995, the tau neutrino in 2000, and the Higgs boson in 2012.[77]

The efforts in unification of electromagnetism and weak interactions into electroweak force has led to discovery of **W** and **Z** bosons by the Italian physicist Rubbia and the Dutch physicist Simon. The weak interaction is mediated by the exchange particles $W^{+/-}$ and Z^0 which are called intermediate vector bosons. The W is involved in commonly observed processes such as the decay of the neutron, beta decay, the decay of the pion(Fig.5.8 & 5.9).

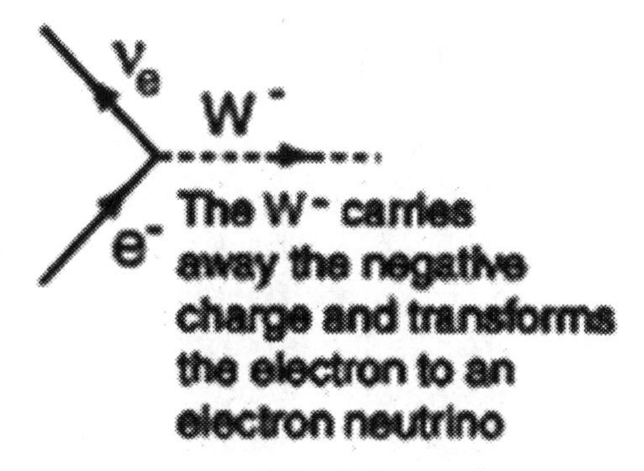

Fig. 5.8

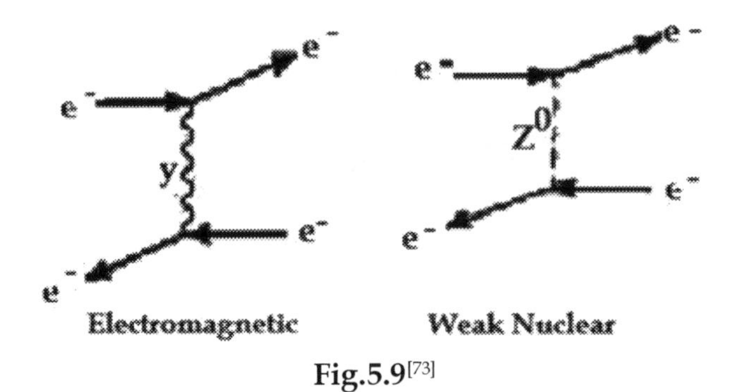

Fig.5.9[73]

We know that the quantum theory is to describe the micro world and the general theory of relativity describes well the macro world. These two theories of modern physics refuse to come to the same platform for handshaking and it is not possible to have a single mathematically compatible model accommodating both. But as the effect of gravity in micro level is negligibly small, standard model successfully explain the interaction of the particles at atomic level. At macro level the effect of gravity dominates and standard model has to remain restricted with the three fundamental forces only. The interaction of subatomic particles are through the four fundamental forces. Richard Feynman, the extraordinary mathematician and physicist, and probably one of the greatest science lecturers of all time, invented a simple way to calculate the probability of collisions, annihilations, or decays of particles, which is called Feynman diagrams(See Appendix-C).

Higgs Boson

Standard model of particle physics predicted the existence of Higgs Boson which is a theoretical, elementary subatomic particle. British physicist Higgs predicted Higgs particle in 1960 which is commonly called Higgs bosons, and thought that this particle was responsible for mass of the other particles. It has been thought that without this τγparticle there will be no gravity and no universe. While discussing about Higgs boson, we should consider three things—the Higgs field, the Higgs boson and the Higgs mechanism.

The Higgs field is a quantum field that the Standard Model of physics predicts and it pervades all space of the universe and interacts with particles. The Higgs field is different from other fundamental or basic fields of nature—such as the electromagnetic field. First, it is a scalar field—i.e., it has magnitude but no direction. Unlike the carriers of the fundamental force fields which are having spin, the Higgs field, the Higgs boson, has zero spin. Second, the Higgs field has the unusual property that its energy is higher when the field is zero than when it is nonzero. It is presumed therefore that when the universe got cooled and became less energetic in the aftermath of the big bang, the elementary particles which were massless, acquired their masses through interactions with a nonzero Higgs field. The elementary subatomic particles have variety of masses because different particles have different strengths of interaction with the Higgs field[87].

The Higgs mechanism has an important role in explaining electroweak theory, which combines interactions through the weak nuclear force and the electromagnetic force. It clarifies why the carriers of the weak force, the W and Z particles are heavy while the photon which is the carrier of the electromagnetic force, has a mass of zero. A direct indication for the existence of the Higgs field is the experimental verification of existence of Higgs boson.[87]. The Higgs mechanism in particle physics is a type of mechanism generating mass for the elementary particles. The theory proposes that the particles gain mass by interacting with the Higgs field. The Higgs boson is the subatomic particle that acts as intermediary between Higgs field and other particles. W and Z bosons and fermions are having masses because they interact with Higgs boson whereas photons, gluons do not interact directly with Higgs boson and hence are massless. The stronger is the interaction of the particle with Higgs boson, the more mass it will gain and will have more inertia.

We know that proton is composed of quarks which are held together by gluons. When the proton is acclerated at 99.9999991% of velocity of light, quarks and gluons collide and explode with huge energy which gives birth to Higgs particle. This Higgs particle has 100 to 200 times the mass of a proton and lasts less than millionth of a billionth of a billionth of a second before decay into a spray of other particles.

For many years the scientific community wanted to know how the Higgs mechanism was functioning. As the accelerator of generating so much energy for the proton acceleration was not available, it was not possible to verify Higgs prediction. In 1993, Miller, a physicist at London attempted to draw a picture of the Higgs mechanism by drawing analogy with some event as illustrated below.

Let us assume that somebody from a balcony is looking down to watch a party. In the big hall, there is sufficient space and the guests are evenly spaced in the hall. They are trying to move from one end of the hall to the other end and are walking freely and no body is coming in the way of their movement. Here the hall represents the Higgs field, and the guests are the Higgs bosons. Then a celebrity enters the hall and all the guests surround the celebrity and press so hard that his movement gets struck. Then when he moves, the whole crowd start to move along with him in such a way that it is difficult to stop that movement of the crowd. The whole clump acquires some inertia of mass. Here the celebrity is analogous to a particle of matter. At that time another celebrity enters the hall but the crowd is so busy with the first one that the second celebrity passes through the hall quickly and without drawing any attention of the crowd. Thus this second celebrity behaves like a massless and weakly interacting particle without causing any disturbance to the Higgs field and therefore behaves like photon or other massless particles.

Because of the nature of the Higgs field and peculiarity of Higgs boson by virtue of its mass and extremely fast decay it was really difficult to measure an individual Higgs boson. Very recently, with the application of enough force and energy in LHC(large hadron collider) it has been possible to knock some bosons out of the Higgs field in a state which can be further studied.

In our discussions of the earlier paragraphs we have seen the different elementary particles and how they interact in the framework of the standard model of the particle physics. Fig.5.10 shows seventeen such particles placed in three separate platforms on the basis of their nature.

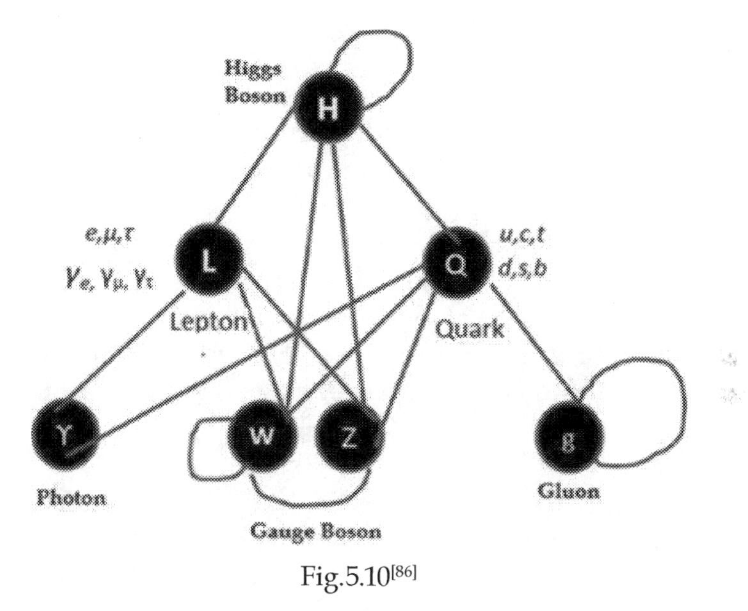

Fig.5.10[86]

Unification of fundamental forces

It was 1960, when three physicists, Glashow, Weinberg and Salam using standard model of particle physics and group theory, proposed a mathematical model to forecast a universal force, electroweak force, combining weak force and the electromagnetic force. Thereafter attempt was made to combine the other force, strong nuclear force with the above two forces to form a Grand Unified Theory(GUT). It was predicted that at some enormous energy of 10^{15} Gev, the srong nuclear force becomes similar to or unified with the electroweak force. In fact, this was not really grand as it did not include another important fundamental force of nature, the gravity. Also it lacked prediction of a number of parameters out of the theory proposed. However this was one sincere effort to get a unified theory. The presumptions are : the strong nulear force gets weaker at higher energy whereas the electroweak force gets stronger at higher energy. At a very high energy level called grand unification energy, three forces supposed to have same strength and appear as different aspects of a single force. It is also predicted that at this energy, the different particles having half integer spin like electrons, quarks are supposed to behave as same. This is another sort of unification and that is at the particle level.

The approximate value of the grand unification energy is imagined to be more than 10^{15} GeV. This scale of energy is close to the magnitude below the Planck scale and thus not within reach of man-made earth bound colliders. With the present generation of particle accelerators, collission at about a hundred GeV is achievable and further upgradation of the order of a few thousand GeV may be possible in the future machines. Therefore it is impossible to achieve the grand unification energy at laboratory to verify experimentally the theory.

Grand unified theories do not include the force of gravity because of the problem in combining the general relativity, a classical theory, with the uncertainty principle of quantum mechanics. But this unification is not so important at micro level when we are dealing with elementary particles because force of gravity appears as such a weak force that its effect can be neglected.

Supergravity theory was suggested in 1976 to combine integer spin perticles with half integer spin particles to predict superparticle unifying the matter particles with the force carrying particles but it was not popularly accepted.

The physicist Paul Dirac developed the first string theory in 1950 where instead of a single point in space, the elementary particle was considered as vibrating string (open or closed) having a length but no other dimension like an infinitely thin piece of string. Further discussions on string theory has been made in the chapter of 'Creation of universe' and Appendix-B.

CHAPTER VI

VEDANTA AND QUANTUM MECHANICS

At the beginning of the 20th century, when the scientific community was feeling complacent with the powerful laws already established by Newton, explaining everything from the motion of the planets to the motion of small objects, human civilization was developing with amazing machines, engines, tools and various instruments, a few adventurous scientific minds decided to cross some new and unknown frontier beyond the range of Newtonian universe. The path finders were Heisenberg and then Schrodinger. This was, to some extent, necessary because when the classical physics entered into the tiny world of atoms and subatomic domain, Newtonian justifications started failing to prove the experimental results. Even at the beginning of this new journey to the wonderland, no body, including the two pioneers, could imagine that this would usher a revolution in the study of physics in the coming days[90][91].

The first revolutionary physicist Heisenberg, comparing his journey to the new world of physics with that of Columbus, dumped the Newtonian universe and entered into hinterland of quantum physics. Heisenberg was confident in his conviction and commented "'it is impossible to open up new territory unless one is prepared to leave the safe anchorage of established doctrine and run the risk of a hazardous leap forward.", Heisenberg discovered a new world like Columbus discovered America, where his abstract matrix mechanics propounded in 1925 could explain the behaviour of the atoms which was so far beyond the ability of the Newtonian universe to vindicate[91].

On entering the realm of quantum physics, the scientists came to know that the world is not ultimately understandable by way of materialism. On this matter, Heisenberg said—"materialism rested upon the illusion that the kind of existence, the direct 'actuality' of the world around us, can be extrapolated into the atomic range." He also said that "the naive materialistic way of thinking is an obstacle to understanding the quantum concept of reality."[91]

With the hypothesis of wave-particle dualism Schrodinger was inspired and got himself involved in formalising the hypothesis in terms of mathematical formulae and finally arrived at Schrodinger's wave function. Although Heisenberg and Schrödinger both started their desperate journeys in the quantum realm at about the same time, Heisenberg took a short-cut method and arrived at solution earlier. Heisenberg started from Bohr's model straight and plunged into the quantum world. He put all the possible electron jumps within the electron orbits of the atom into a big table, called a matrix. He was then able to find out proper laws for these matrices, and directly land on the quantum realm with this matrix mechanics. Schrodinger and Heisenberg found that they both had arrived at the same solution—the wave mechanics and the matrix mechanics were mathematical variations of the same quantum mechanics[91].

Heisenberg and Schrödinger, both deserved the credit of establishing a strong footing on mathematical platform for quantum mechanics. But it was Niels Bohr who could successfully built a bridge between the two conceptions—"matter waves" described by the Schrodinger's wave functions and the idea of matter resting on nothing but the probabilities in Heisenberg's matrices. Bohr's proposal was a strange combination of both—the particle as well as wave nature of the world. The two views are complementary, no single view by itself telling the whole story[91].

The matter wave is not ordinary wave like sound wave or waves in water which are propagated due to vibrations in the physical medium. The matter wave or quantum waves are nonmaterial wave of probability and not actually describe the physical properties of the particles but only their probable or potential properties.

Therefore, if we look into the electron orbits in an atom, the orbits are not the physical path followed by the electron but rather a wave of possibility for the electrons to be found in different locations. Therefore, instead of applying Newton's law of motion, the quantum laws should be applied for describing the movement of these waves potentiality. The existence of particles is gone—only remain the possibilities of their existence. Thus the physical presence of substance of materialism has reincarnated as wave functions, describing particle at a particular location as the probabilities for it to be in that location. It appears that as we dip into the realm of quantum mechanics, the apparent world of hard matter based on solid material particles so far dealt in classical physics dissolve in the airy cloud of nonphysical probabilities. Entering further into the quantum world, things were appearing very peculiar. With the evaporation of matter, determinism, and finally separatibility, the physicists observed quite unexpected behaviour of the quantum world in their observations and when explored further in the quantum realm, very odd things were coming up beyond the common level of understanding that created puzzles[91].

Being tempted with such observations Richard Feynman remarked, "If you say that you understand quantum mechanics, and if you are NOT surprised, then you really did not understand quantum mechanics !" The reason is that the subject is full of mysteries and puzzles and anybody having a first hand taste of the subject may always be tempted to dub the subject as the most supernatural and unscientific one. Feynman also added, "I can safely say that nobody understands quantum mechanics."[92]

An identical situation has been described by Upanisads where the conception dominating is that Brahman cannot be understood by reasoning. If it can be understood it is not Brahman. This has been elaborated in a fine story of two Yogis. One used to meditate all the day and the other one used to read the spiritual scriptures all the day. At the end of each day, the second one used to say," I do not understand, I am not able to understand." One day with utter surprise, the first Yogi observed that the second one was shouting, "Now I understand, I understand. "The first one asked him "What has happened? After so many days, how have you understood everything suddenly?" The second Yogi replied confidently "I now understand that it is not possible to understand this."[106]

Sir Roger Penrose, the well known physicist and mathematician, said "Quantum mechanics gives us wonderful predictions and experimental confirmations for small-scale scenarios, but it gives us nonsense at ordinary scales."

Adding fuel to the fire, some of the quantum mechanical facts appear to be really very difficult to digest with the level of understanding of a common man. For example, we know that electron is a quantum particle like photon. In contrast to its particle character, it has been proved that the electron moves through two openings simultaneously!

Double slit experiment

As we have witnessed the wave nature of light from the result of the double slit experiment in the college laboratory with a beam of monochromatic light projected through two narrow holes in a screen, the light spreading out from the two holes interferes, just like ripples interfering on the surface of a pond, to produce a characteristic pattern on the screen behind.

Now the mystery is that light is also described as a stream of particles, called photons. In the same type of double slits experiment, if we send stream of photons in such a way that individual photons can pass through the slits and arrive at the screen behind which is a photoelectric detector, we observe that they still create an interference pattern, indicating as if each photon passes through both holes and interferes with itself on the way through the experiment. Quantum theory has tried to give an explanation to this phenomenon by saying that, even if we consider that light is a stream of photons, there is no mechanism to find out, even in principle, which photon has gone through which slit. The "indeterminacy" allows fringes to appear.

Chiao and his colleagues extended the scope of the experiment further by putting two separate polarizing filters, one with left handed circular polarization and the other with right handed circular polarization in between the light source and the individual slit and found that the interference pattern on the back screen vanished. By using the filters, in principle, it was possible to identify through which slit photon stream were passing and created impression on the back screen though the indeterminacy in the movement of photons through the slits still remained. But when a third polarizing filter, call it the erasing filter, is placed in between the slits and the back screen to scramble or erase the effect of the earlier two filters, it was again not possible to say through which slit a particular photon had passed but this time the interference pattern reappeared[93][97].

All of these experiments were carried out using beams of individual photons, and by using classical physics we cannot explain in what way such result has been arrived. Therefore the result of the experiments predicts that the interference depends on "single photons" going through both slits at the same time but undetected. The question remains that how does a single photon know how it should behave in order to match the presence or absence of the erasing filter on the other side of the slits? Or starting out of its journey how it behaves in a different way for different experimental set up as if it is aware in advance the nature of test it has to undergo ? The incidence can be described as—If you observe the movement of the photon, it starts behaving as a particle and when the movement of photon is not tracked and the result is seen, it behaves as a wave!Heisenberg said, "The path [of the photon] comes into existence only when we observe it."

Non local interaction

There is another puzzle called the 'non-local' interaction. Suppose two quantum particles have been together like twins. Now separate them and take to very far away place from each other so that they don't see or talk to each other. If now some experiment is done on one of the particles, we shall see the other particle which is far away, reacting to this experiment on its twin immediately! This fact was theoretically described by Bell of CERN in 1960s and experimentally observed by Aspect in 1980s in Paris. The exploration shows that when two photons are ejected from an atom in opposite directions, the behaviour of these two photons show astonishingly entangled mutually. If the state of one of the photons, so detached, is measured, it instantaneously affects the state of the other one, wherever it may be[93].

Unlike classical physics that distinguish the particles and energy and does not attribute any wave nature to the particles, quantum physics treats the particle as both wave and particle and says that every subatomic particle is associated with a wave function which is a mathematical expression depicting the probability of a particle to be in a particular location or state of motion. From the observations like spectrum of black body radiation, photo electric effect etc. there are postulations that energy exists in discrete quanta which is in contradiction to the observations of classical physics. Also the behaviour of elementary particles are observed highly unpredictable for some tests. For all these reasons, in the first part of twentieth century, quantum mechanics was very much successful in pushing the modern physics to have much insight to the atomic and subatomic activities and provided a boost to the advanced technologies based on the study of the atomic objects to attain a new height. But at the same time, it posed serious challenges to the imaginations of the physicists. A trend to take the help of metaphysical interpretations became immanent to account for the violations noticed in the quantum mechanics.

In the years 1924-1927, Bohr, Heisenberg and others formulated the Copenhagen Interpretation which was to explain the experiments and its mathematical formulations. While doing this, they argued beyond the world of empirical experiments and pragmatically predicts to explain the violation noticed. According to this interpretation, the act of measurement causes the calculated set of probabilities to 'Collapse' to the value defined by the experiment. This feature of mathematical representation is called 'Wave Function Collapse'.

The physical theories in general are thoughts, depicting a reality that is beyond our direct experience. We feel our spontaneous sensations of vision, sound, touch etc., and never directly encounter the abstractions of "atoms" or "fields" which are only indirectly inferred from experience. We never really experience those useful abstractions directly and are not sure whether they really exist the way we imagine. Science tries to balance the belief in some objective reality with the fact that we can never know the subject in itself. As Heisenberg wrote, "We have to remember that what we observe is not nature in itself but nature exposed to our method of questioning".

A theory of quantum mechanics called "indeterminacy" says that a particle can be in two states at the same time mathematically. Schrodinger did not accept this and wanted to show that it was not true, so he came up with the proposal of an experiment which was only hypothetical, and could not be really done.

Schrodinger's cat

In the early 1930's Schrödinger published a way of thinking about the circumstance of radioactive decay that is still a useful concept. The thought experiment is as follows:

Let us imagine an apparatus containing just one Nitrogen-13 atom (half life of radioactive Nitrogen-13 is 10 minutes) and a detector that will respond when the atom decays. Connected to the detector is a relay which is again connected to a hammer, and when the Nitogen-13 atom decays the relay releases the hammer which then falls on a glass vial containing poisonous gas. We take the entire apparatus and put the same in a box. We place a cat also in the box, close the lid, and wait 10 minutes. After 10 minutes we ask: Is the cat alive or dead? The answer according to quantum mechanics is that it is 50% dead and 50% alive. This is because probability of decay of that atom of Nitrogen-13 in 10 minutes is 50%. According to quantum mechanics, since we can't see inside of the box to know if the cat is alive or dead. As per Copenhagen interpretation, the cat is both alive and dead. It exists in a state of 'superposition'[94] (see Appendix-D)

Spooky Action

The two entangled photons originated from pion decay are fired in opposite directions and let one of these photons flies to right. If we measure its spin in X axis with precision (it is possible), its Y axis spin cannot be measured as we already know the X axis spin as per quantum mechanics. Let us then go to the second photon which flew left. Being entangled photons, we already know its X axis spin which will be exactly opposite to the first photon. Now the question is : Can we measure with absolute certainty the Y axis spin of the second photon though we know its X axis spin without measuring it?

Einstein said : "Yes ". How would the second photon know that you measured the first photon? Relativity says that the "knowledge" of the measurement on the first photon can only travel at the speed of light.[100]

Quantum mechanics says: "No." You can't measure the y-axis spin with absolute certainty. It doesn't matter whether the two photons were separated by a centimetre or ten Kilometres. The very instant you measure the x-axis spin of the first photon, the y-axis spin of the second photon becomes impossible to measure! The "knowledge" of the measurement is instantaneous as they are entangled. Einstein called it "spooky action at a distance".

To explain this paradox, some scientists thought that the "hidden variables" exist in the photons that allow them to behave this way. For entangled photons the hidden variables are the same. Hidden variables are variables that are yet to be discovered [100].

Our idea about object having definite properties whether we are observing it or not is the basic requirement of realism which goes when we enter into quantum realm. Quantum theory also defies the principle of locality which restricts the communication faster than the speed of light as already discussed. The physicist John Bell showed that all physical theories that obey realism and locality support the numerical inequality produced on a particular combination of measurements performed on identically prepared pair of particles. He also showed that this inequality is violated for entangled particle pairs as predicted by quantum physics.

Simple explanations of the results of the twin particles experiments and to the entanglement problem have been tried. Some physicists are of the opinion that there is no separation at all. The twin particles are interconnected with each other as all particles in the universe are. The information does not need to pass faster than the speed of light because actually the two particles are not detached. In other words, it is opined that the twin-electrons experiment actually proves that the particles in the universe are not detached at all[105].

Limitations of western science

So far, the modern physics has followed a path with reductionist view of the universe that started from seventeenth century through the new scientific revolution created by Newton. First the concept of aether which was thought by Maxwell to be necessary for propagation of electromagnetic waves at the speed of light, was negated and thereby the basis of the beliefs in the metaphysical world, that all the objects in the universe are floating on the universal cosmic fluid whose vibrations connect all with one another and rest of the universe, was shattered. The next jolt on the conception of universal consciousness came from the experimental assertion by the physicists that matter and light were particles, electricity is the flow of electrons, the charged particles and the conception of universal continuity was disturbed. The atomic theory and the quantum mechanics brought the idea of quanta, the discrete micro entity existing in everything even in space and time[101].

It is surprising that a major population of quantum physicists have always been searching the answers to these mysteries in the ancient Indian Vedic texts because so many mysteries identical to that happening in quantum mechanics could find explanations in these ancient scriptures. The basic reason for this is that until recently, western science restricted its domain to empirical investigations of the objective world and there was no universally accepted, comprehensive theory of consciousness existing in it whereas consciousness is inherently subjective and in our lives, consciousness and its natural consequences are fundamental to every thought, word and action. Consciousness is the most pragmatic basis of our existence and without a practical theory of consciousness, scientific explorations cannot be complete. Therefore without consciousness, quantum mechanics also becomes merely mathematical theories of chance and probability to explain observations at subatomic level.

The quantum physicist Schrodinger said, "This life of yours which you are living is not merely a piece of this entire existence, but in a certain sense the whole . . ." He continued, "This, as we know, is what the Brahmins express in that sacred, mystic formula which is yet really so simple and so clear; *tat tvamasi,* this is you. Or, again, in such words as I am in the east and the west, I am above and below, I am this entire world."

'Tat TvamAsi'—that thou art is the great message of Self-revelation of the Upanishads:. The nature of the inner essence of oneself and the nature of Brahman is one and the same infinite consciousness, free from all change, formless and all-pervasive. The ancient sages realized with all spiritual experiences of the their lives, that you are not separate from Brahman: you are no other than the Brahman Itself and you are the Existence-Knowledge-Bliss Absolute. The individual and the cosmic, *Avidya* and *Maya, Jiva*and *Isvara, Atman* and *Brahman* are the aspects which correspond themselves to the meaning of the terms 'Tat' and 'Tvam', 'That' and 'Thou'[102].

There was a clear continuity between Schrodinger's understanding of Vedanta and his research. Schrodinger read exhaustively and thought and assimilated deeply the techniques of ancient Hindu scriptures and rewritten them into his own words and ultimately started to believe in them. This was evident from many of his writings. Walter Moore, the biographer of Schrodinger has nicely narrated the state of mind of Schrodinger and quoted, "The unity and continuity of Vedanta are reflected in the unity and continuity of wave mechanics. In 1925, the worldview of physics was a model of a great machine composed of separable interacting material particles. During the next few years, Schrodinger and Heisenberg and their followers created a universe based on superimposed inseparable waves of probability amplitudes. This new view would be entirely consistent with the Vedantic concept of All in One".

Schrodinger wrote in July 1918, "Nirvana is a state of pure blissful knowledge . . . . It has nothing to do with individual. The ego or its separation is an illusion. Indeed in a certain sense two "I"'s are identical namely when one disregards all special his Karma and to develop it further . . . . when man dies his Karma lives and creates for itself another carrier."

Throughout his life, Schrodinger maintained his belief in Vedanta which he expressed through his elegant essays. He, definitely, grasped the theory of Vedanta as 'Mahavit' as described in Chhandogya Upanisad but could not achieve a practical realization of the absolute truth and thereby a state of 'Atmavit', by realizing the 'Atman'. Schrodinger in his book 'My World View' wrote "Vedanta teaches that consciousness is singular, all happenings are played out in one universal consciousness and there is no multiplicity of selves".

The present day ambition of the scientists is to have a unified theory which will manifest the four fundamental forces, the strong and weak forces, electromagnetic force and the force of gravity. The physicists consider that at the time of big bang, the unified field was existing for one trillion, trillion, trillionth of a second and this was the singularity. The sages of Upanisads called it Brahman.

The Vedanta and the quantum physics deal with the same philosophical subject—the absolute reality and its immeasurable, inconceivable and intrinsic nature. Though the approach of the two different disciplines are quite different, carry divergent points of view and treat the subject with widely different language, it is possible to reconcile the conceived ideas, because of the commonality of the subject.

Science is nothing but one of the ways of humanity's quest for absolute truth and it has never been the ultimate acquisition of absolute truth. Since scientific truth is constantly changing, it cannot be absolute truth and modern science must contain a mixture of truths and non-truths. We can imagine that someday science will overcome these limitations and arrive at absolute truth.

Sri Aurobindo, in his book, the Valley of the False Glimmer said, "Science has missed something essential; it has seen and scrutinised what has happened and in a way how it has happened, but it has shut its eyes to something that made this impossible possible, something it is there to express. There is no fundamental significance in things if you miss the Divine Reality; for you remain embedded in a huge surface crust of manageable and utilisable appearance. It is the magic of the Magician you are trying to analyse but only when you enter into the consciousness of the Magician himself can you begin to experience the true organisation, significance and circles of the Lila."

We have already seen that the theoretical dynamics of matter accounted by the four fundamental forces do not apparently suggest any unified field theory, through their interactions with each other, which can explain the natural ordering of the universe. The scientists are of the opinion now that a so far unidentified fifth field of interaction may help them to bring about the unification. The quest for such a field has lead them to speculation about a sub-quantum reality as described by the ancient sages as Luminiferous Aether, an all pervasive invisible medium. We discussed earlier that existence of such aetheric force was denied by the scientists as they believed that such metaphysical force could not play any significant role in physical interactions. But now, the theoretical physicists are contemplating existence of Quantum Vacuum in the invisible sub-quantum field as the Aetheric force projecting the unidentified fifth field of the nature. The Aetheric void is now being reconsidered as the quantum vacuum which is a highly charged cosmic medium that influences the space-time motion of the observable universe appreciably.[104].

The scientific world is moving towards this new field of thought only now when the ancient sages of Vedic India could visualized these realities long ago and postulated the existence of the five fields, aetheric Plenum or the Fullness being the fifth one from which everything emerged. As described in Atharva Veda, the observable or material universe emerged from a point or seed which was rested at the resonant threshold of that Etheric plenum, called the 'Embryo' or 'Hiranyagarbha'. In scientific cosmological language this point or seed is the singularity from which the universe came into being. In metaphysical terms it is Luminiferous Aether or AUM or the primeval vibration from which the universe originated[104].

The womb of creation as predicted by the ancient sages, that it is not measurable, is also realized by the physicists. The laws of physics break down at this resonant threshold because they are unable to measure its infinite density.

Through proposed theory of quantum vacuum a new chapter of quantum physics is being opened, called as Vacuum Physics with the beliefs that it is the source and sink of all matter and perhaps has given the birth of observable universe. But the discovery of quantum vacuum has not solved any problem except postulating a zero point beyond which the empirical science cannot go.

To reveal the modes of interaction of the Quantum Vacuum, a new system of knowledge required to be achieved to understand Time in a fresh way of thinking as Time is the key which can disclose the secrets of matter. For centuries, Time is being regarded as a derivative of Space, more enthusiastically being called as 'Space-time' but, according to the ancient sages, it is quite reverse; Time is what gives birth to Space. Sri Aurobindo said "Time is that mysterious condition of universal mind which alone makes the ordering of the universe in Space possible."[104]

The Absolute is unbound, indistinguishable, eternal, One and only One in its intrinsic state. It becomes intelligible to itself from unintelligible state and then it becomes dense, becoming audible first and then visible. While doing this it had undergone through a process of contraction that slowed down the vibrations of the infinite consciousness and then gradually solidifying it into a spatial field manifesting the material forms. The Absolute thus reduced from its original state of undifferentiated Unity manifested into Form and Multiplicity.

A new general map is needed to enable navigation when dealing with quantum phenomena, complex systems, evolution, life etc. It is now being envisaged that in order to unveil the secrets of the Transcendent dimension, science must learn to inverse this process of divine alchemy. The philosophy of continued disintegration must be abandoned and a cosmological, or whole systems approach should be taken. The material existence should not be taken as collection of separate objects as observed but to be looked upon as undivided whole where all distinct parts merge and unite to reflect the totality. A major paradigm change may change the rules, raise up new questions, categories and distinctions. May be the old controversies between science and humanities, Western and Eastern thought just disappear. Already the thought processes have started to embrace cosmology, the mother of all sciences to have a paradigm shift in the study of exploration of the order of the universe and thereby finding the absolute truth or to approach towards the concept of holistic universe [107].

APPENDIX-A

Theory of Relativity

To ponder upon the subject, we may get the following few items clarified.

Frame of reference: When we say a body is in motion, we must mention the position of the body at a given point of time. Again to specify the position of the body, we must measure the position relative to some other body or frame of reference. In fact, there is no absolute motion, it is always the relative motion. In the frame of reference, we should have the origin from which the measurement of motion is to be taken and the direction in which the distance from the origin is to be measured.

The Galilean relativity is the spatial relativity and termed in the name of Galileo who gave the conception of relative motion and stated that an observer in motion observed an event differently than a stationary observer for having different spatial coordinate system or frame of reference.

Inertial frames : The inertial frames are non-accelerating frames. The frame in which the Newton's law of inertia is valid is called inertial frame. A frame is an inertial frame which is moving at a constant relative velocity to another inertial frame. If there are two inertial frames moving relative to each other, any one may be considered to be at rest relative to other. The translation of the observation of an observer of inertial frame A to that of inertial frame B is called Galilean transformation. An inertial frame A with coordinates (x, t) and another inertial frame B with coordinates (x′, t′) moving at constant velocity v relative to A, the x, t and x′, t′ are related by the following equation :

$$x' = x - vt \text{ and } t = t'$$

The laws of physics will be same for both. It can therefore be inferred that laws of physics are same for all inertial frames.

Non-inertial frames: Newton's law of inertia may not be correct all the time. Suppose the moving body itself is the origin of the frame of reference. Then the body is always at rest whatever be the force acting on it. Also the motion of the objects in the space shuttle moving in the orbit have no relative motion with respect to the space ship though gravity is the force acting on them. The frames which are moving at accelerated speed with respect to inertial frame is called non-inertial frame and law of inertia does not hold good for this frame. Similarly, Newton's second law of motion also is not valid in non-inertial frame of reference.

Einstein's special theory of relativity : The subject deals with the motion of object when its speed is close to the speed of light. Newton's second law of motion becomes incorrect when the object approaches the speed of light. When the motion of an object, moving at speed close to speed of light, is seen from an inertial frame, the event can be explained by special theory of relativity. When the same is seen from non-inertial frame or motion of the object is in strong gravitational field, it becomes the subject of general theory of relativity. In our daily life on earth, these two theories do not interfere normally unless extremely high precision is needed in calculating some movement ; for example relativistic effects becomes very important in GPS(Global positioning system).

In the later part of nineteenth century, Maxwell propounded his famous electromagnetic theory. The theory estimated the value of speed of light. It states that light is a form of electromagnetic radiation and propagates at the speed of light but this speed is something that depends on the inertial frame from which it is observed i.e. it is not same for all inertial frames. But laws of physics was considered to be invariant irrespective of different inertial frames from which observations are made! To sort out this dilemma, scientists brought the idea of aether and considered that the speed of light predicted by

Maxwell was relative to aether and variation of light due to different inertial frames was not due to laws of physics becoming different for different inertial frames but due to relative motion of the inertial frames with respect to aether.

Michelson-Morley conducted an experiment about hundred years ago to find out the dependence of speed of light on the speed of the observer. To every one's surprise it was observed that the speed of light was independent of observer's motion or the inertial frames, reversing the expectation of the scientists that speed of light would be different in different inertial frames. Einstein was very much impressed by phenomenal success of Maxwell's electromagnetic theory and thought that a new mathematical model was required to establish that Maxwell's laws were same for all inertial frames. He dumped the idea of aether and brought the idea of simultaneity to explain why the speed of light would be constant in vacuum and not dependent on observers' frame of reference.

Einstein changed the idea of absolute time and placed his famous theory of special relativity on the basis of two postulates:

The first one is that the laws of physics are same for all inertial frames i.e. if some experiment is done in a moving car (moving at uniform velocity with respect to ground), the findings will be same as that when done in a lab at rest with respect to ground.

The second one is that speed of light is constant in free space. The result becomes revolutionary. Space becomes entangled with time. He brought the conception of space-time event and the 3-D coordinate system used in depicting a point in space became a 4-D reference frame with time as the fourth component of that space-time frame. For example: Let X invited Y for a cup of tea at a mall at the southern part of the city in a restaurant at 4th floor of the mall and reserved a table in the north-eastern corner of the restaurant. So the meeting point is precisely indicated—north-east corner table, 4th floor restaurant and mall at the southern part of the city. It is like three spatial coordinates to locate a point. But one important information is missing. Meeting at what time? So time also an important component to specify the event and thus it is space-time event.

Einstein replaced the Galilean transformation to Lorentz transformation which is expressed by following equations:

$$x' = \frac{x - vt}{\sqrt{1 - \frac{v^2}{c^2}}}, \qquad t' = \frac{t - \frac{v}{c^2}x}{\sqrt{1 - \frac{v^2}{c^2}}}.$$

From the above expression of Lorentz transformation, we find that when the relative speed v between the two inertial frames is too small compared to speed of light i.e. v/c is negligibly small, Lorentz transformation approximates to Galilean transformation.

The idea of simultaneity may be explained by the following three observations with the train and glowing bulb example. A light bulb is fixed at the middle of the train—train is moving and bulb is glowing(fig. A.1).

Observation 1 : The observer is in train. The observer will see that light is reaching to the front and back of the train at the same time.

Observation 2: The observer is standing on the ground. The observer will see that the light is reaching to the back of the train prior to light reaching the front of train.

Observation 3: The observer is moving ahead of train with speed more than the speed of train. He will see that light from the glowing bulb is reaching earlier to the front of the train than the back.

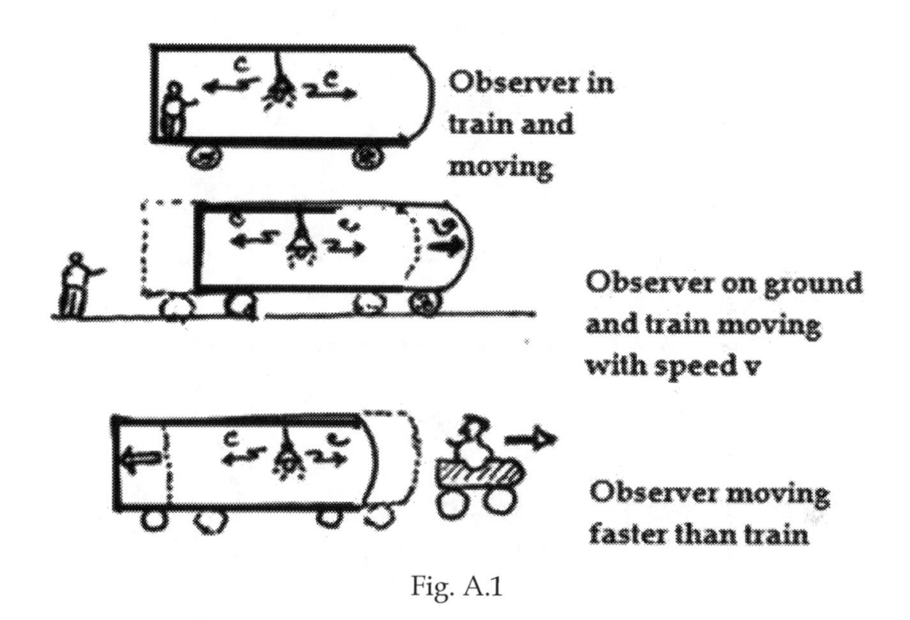

Fig. A.1

Thus we see that the observations or concept of simultaneity and happening of incidences before or after, vary with the positions of the observer i.e. the inertial frames of observations as in all inertial frames the speed of light is the same.

We can discuss concept of cause and its effect which is termed as causality. If an event A is cause and other event B is its effect, then A must occur before B. But like the above example, if some observer finds A occurring before B and other observes A occurring after B, there arises the contradiction! The Einstein's paradox can be explained in the following example:

A moving train carries B at its front part and C at its rear part. A and D are standing on ground(fig. A.2). As soon as train passes by A, a message is sent by A to B who in turn sends the message to C available at the rear part of the train through an instantaneous communication device (obviously with speed much more than that of light). As C passes by D, the same message is relayed to D and D sends the message to A through the same type of instantaneous communication device. If the train moves faster than speed of light i.e. v> c, it will be seen that A receives the message from D before it sends it to B. This is happening because the train is considered to have travelled faster than light. This cannot happen in reality! Thus nothing can travel faster than light. In fact, Einstein's relativity theory has put a limit to the speed of any object and that is the speed of light.

Let us consider another term 'time dilation' which states that clocks at different inertial frames run at different rates. Two observers A

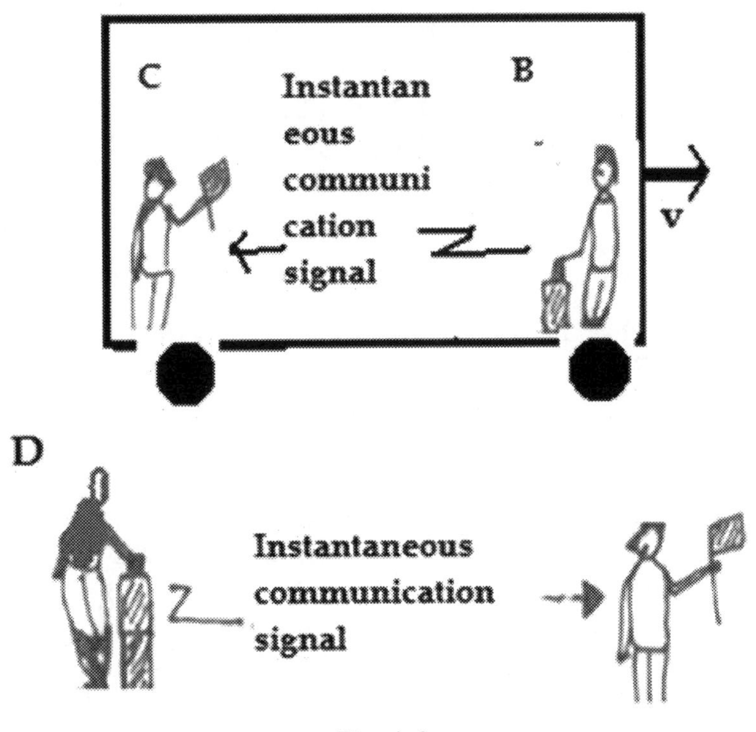

Fig. A.2

and B in two different inertial frames synchronised their clocks when they are stationary. The one observer B starts moving at uniform velocity relative to A. After a time T in the stationary frame, the stationary observer compares his stationary clock with the moving clock of the moving observer. The time in the moving frame is not same as that of stationary frame. The reverse is also true when the moving observer compares his moving clock with the stationary clock. If time in stationary frame at the time of comparison by stationary observer is T and that in moving frame at the same time is T′ then

$$T = \frac{T'}{\sqrt{1 - \frac{v^2}{c^2}}}.$$

When v is very much less than c, T and T′ is approximately same. But as v increases and approaches c, the moving clock runs slower i.e. T′ is less than T—this is termed as time dilation.

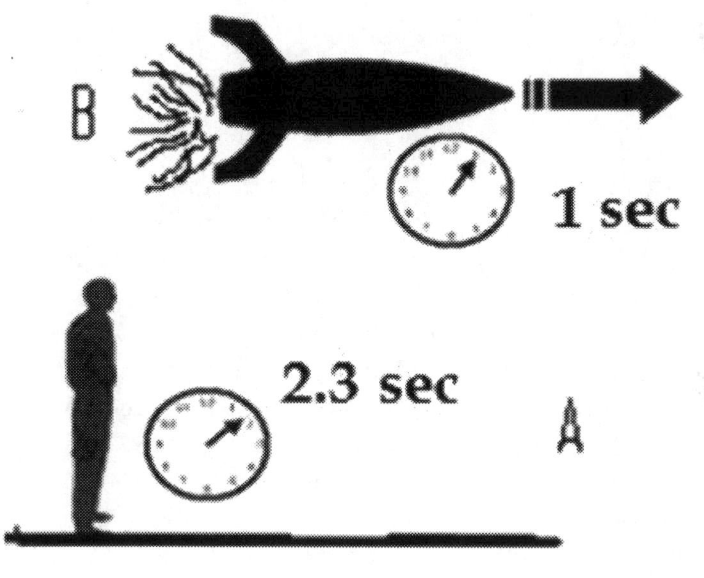

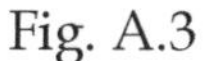
Fig. A.3

Let the speed of B or the rocket by which he is moving is 9/10 of c. Then 1 sec in the rocket will be 2.3 sec in the clock with A who is at rest. As the speed of rocket increases and approaches c i.e., the spatial speed increases, the time speed i.e which is the time observed by B in the rocket relative to the time observed by A goes down(fig. A.3).

Einstein's theory of General Relativity: Everybody knows that if anything is dropped, it falls down straight and does not go up. This is due to gravity and the force acting on it is gravitation. Newton presented his law of gravitation by stating that the gravitational influence or the force of gravity between two masses M and M′ separated by a distance d is proportional to the product of two masses i.e. M * M′ and inversely proportional to square of the distance between the point masses i.e. d^2 (fig, A.4). Therefore,

Force of gravity ∞ M * M′ / d^2

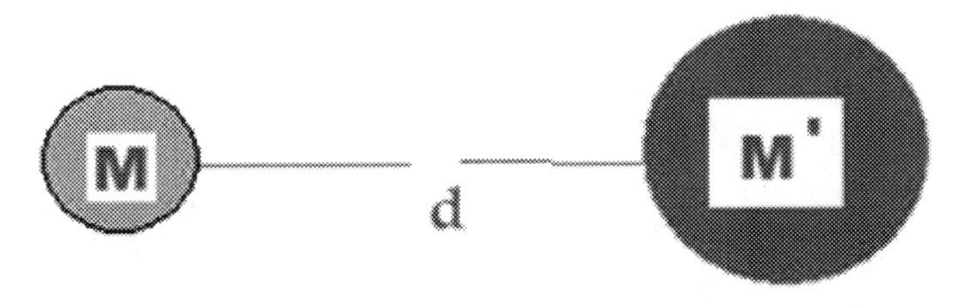

Fig. A.4

Newton's gravitational law assumed that the gravitational influence travels instantaneously. It means that if the sun is suddenly vanished due to a black hole passing by, its gravitational influence will reach earth instantaneously. But this will not actually happen. Afterwards it has been clear that such influence will travel with finite speed and the effect of such incidence will be detected by earth after 8 minutes which is the time taken by light to reach earth from sun.

In Newton's laws there are conceptions of two types of masses—inertial mass in the law of motion and gravitational mass in the law of gravity. Though Newton was aware that these two masses are of same value, he could not have idea of the reason for this.

Principle of equivalence:

Let us compare the motion of an object due to free fall in the gravitational field on the surface of earth registering acceleration due to force of gravity and the object that moves in deep space in a spacecraft far away from any gravitational influence and with same acceleration.

As per Newton's law of motion: Force required to move an object of inertial mass m_i and impart a uniform acceleration 'a' will be

$$F=m_i a$$

If the gravitational mass of the same object is m_g and is allowed to fall freely with acceleration 'a' under the gravitational force of earth on m_g, then force due to gravity will be

$$m_i a= - G m_g M_g / R^2 = g m_g$$

G is the gravitational constant, $g = -GM/R^2$ is the gravitational field, R is the radius of earth and M_g is gravitational mass of earth, the negative sign is to indicate downward movement of the object. Therefore as m_i and m_g are equivalent and hence same :

Inertial acceleration(a)= gravitational field (g)

The equivalence principle says that a uniform acceleration is equivalent to uniform gravitational field.

Einstein, therefore reasoned that all the experiments conducted in a free falling local frame of reference where effect of gravity is absent, will give identical results when conducted in deep space where there will be no gravitational influence. Here 'local' represents a small region of space where gravitational field is uniform. He opined that the gravity is not some type of force that pulls the things as we considered, but was an event related to free motion in space-time or curvature of space-time. The curvature of space and time is caused by the presence of a massive object (say Earth) nearby. When some object moves along and passes by the massive object, it appears as being pulled towards the massive body, but actually it is not like that. The object moves along the same straight line as it is considered to do in empty space. This straight line looks as if it is curved.

Einstein utilised this equivalence principle in explaining gravity under the framework of his theory of relativity. Question arises whether the property of an accelerated system is due to gravity or change of velocity. The answer remains in the frame of reference from which you are observing this. Let us consider the thought experiment :

Let X is inside an elevator which is at rest, relative to earth's gravitational field(fig. A.5). X is also at rest and due to gravitational force on his body, he pushed the floor of the elevator downwards. Since he can go neither down the floor of the elevator nor going out in the air, the floor also pushes him upwards and due to this reaction force X feels his weight. In the elevator, if allows a ball to fall, it will fall straight due to gravity. Suddenly, due to some accident, the elevator starts falling freely from some height due to snapping of the support cable. In such condition, floor of the elevator cannot push X upwards. X remains at rest relative to elevator since both X and elevator fall freely at equal acceleration towards the centre of the earth due to gravity, the reaction force on X vanishes and X starts feeling weightless. If he lets the ball to fall now, he will see the ball floating right next to him. He will see that there is no gravity inside the elevator. This shows that effect of gravity disappears when the frame of reference is in state of free fall.

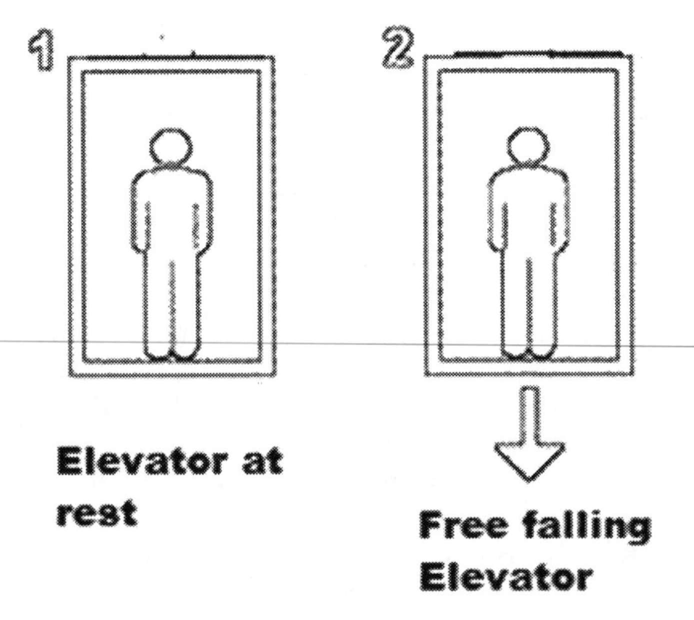

Fig. A.5

Let there be another thought experiment. Suppose a car is at rest and a man standing inside the car, leaves a ball to fall (fig A.6a). The ball will fall downwards straight due to effect of gravity and the man becomes sure about the flat space inside the car. Now the man enters into another car which starts moving with uniform acceleration without the knowledge of the man(fig A.6b). If the man release the ball again, he will see the ball is falling down following a curved path! Since the man is not aware of the motion, he will think that there is some disturbance in the car and the space inside the car is curved. But to the man the car remains as it was earlier and he knows that physical laws in all inertial frames are same. So the car must be in motion. Now, this motion is detected by him without reference to any other object ; hence he thinks that it is absolute motion. But no motion is absolute in reality. So the man infers that gravitation has resulted this curved space inside the car[109]. So uniform inertial acceleration of the car has created the event equivalent to gravitation.

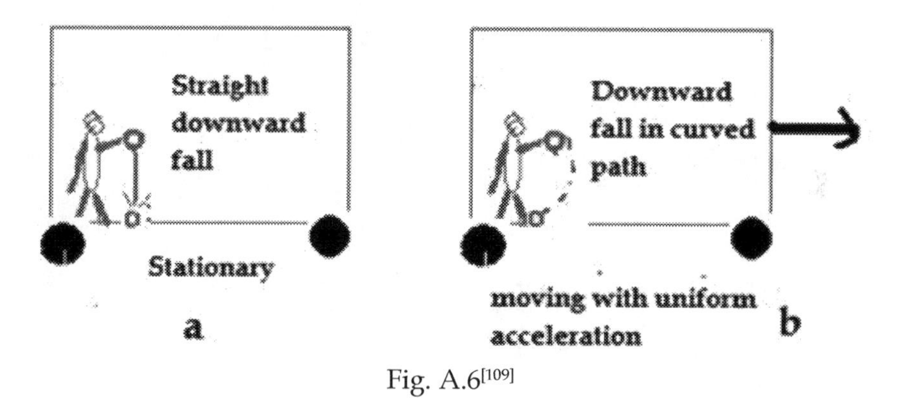

Fig. A.6[109]

Space-time curvature: Let us have an experiment. A bullet is sought from a gun (as shown in fig. A.7) which traverses initially parallel to earth's surface and then follows a trajectory to fall ultimately at a distance from where it is shot. A ball is also thrown in the same direction, of course this time with less force and the ball touches the ground after some time and at a distance much less than the bullet.

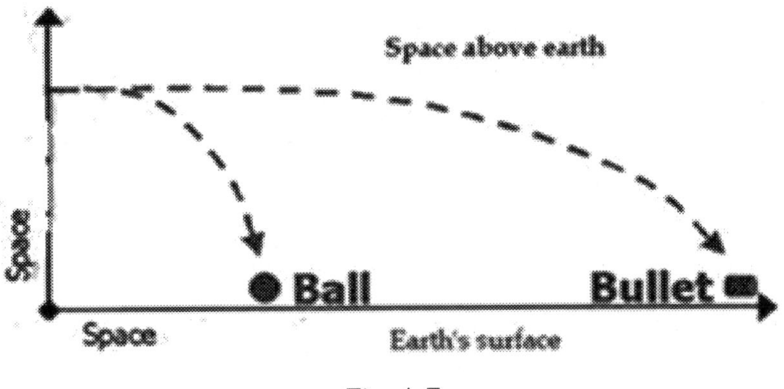

Fig. A.7

A question may arise that if the space above earth is curved due to gravitational influence of earth, the bullet and the ball should have followed the natural curved path of space and reach same distance. Then why is this not so ? Suppose A and B starts from same location on the curved surface of earth towards north following straight lines. Both follows same path whatever be their speed and ultimately reach same location. Why is not this happening for the bullet and the ball? The matter will be clear if we plot their motions in space and time frame as below:

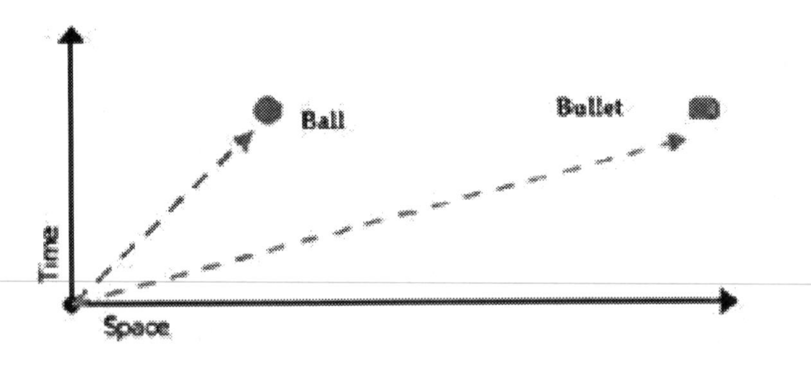

Fig. A.8

Einstein explained that gravity is not the curvature of space alone; rather it is curvature of space-time. When the motion of the bullet and the ball is described as movement in space-time frame as above, it is seen that they are on different tracks in space-time frame, the bullet traverses more space than the ball, having more initial velocity(fig. A.8). Their movement in space domain is visible but movement in space-time domain can be imagined only. If now two bullets of same material or bullets of different materials and hence of same masses are fired by the gun with same initial velocities in same direction, they reach at same location following same path. Therefore the path that is followed by an object in space-time, does not depend on the mass of the object or material of the object. It depends only on the initial direction and the velocity of the object.

Riemannian Geometry :

Let us take a flat piece of paper. If we draw a number of points on the paper and between any pair of points, if we draw a line, that will be the shortest path between those two points. Also if we draw another line parallel to the first one, those two line never meet with each other. The well-known Pythagoras theorem, trigonometry etc. are based on the geometry of flat space, called Euclidean geometry.

Now if we take a curved piece of paper, say in form of a cylinder, or a globe in form of a sphere, the shortest curve between any pair of points on such a curved surface is called a minimal geodesic. We can form a minimal geodesic between two points by stretching a rubber band between them. We may notice that sometimes there are more than one minimal geodesic between two points. Take a globe, you will find that there are many minimal geodesics between the north and south poles of the globe. We cannot draw lines on the sphere other than the minimal geodesics. There are no lines on a sphere! These will be the curves only which we see in Euclidean space and these wrap around the sphere and cannot be extended. On a cylinder, most of the minimal geodesics wrap around the cylinder and cannot be extended but some can be extended to lines. Still we can estimate the length of the hypotenuse of a triangle. Also, the circumference of a circle and the area inside the circle can be estimated. It is the Riemannian Geometry that deals with the curved surfaces. The amount of curvature of space can be estimated by using theorems from Riemannian Geometry. Einstein, studied Riemannian Geometry before he developed his theories.

APPENDIX-B

Inflation and scalar field: Mathematically the energy associated with the Scalar field shows the characteristics that with very high energy density, the decrease in density, with respect to expansion of universe, is very slow. When the density attains certain value, it stops behaving such way and follow the pattern as followed by ordinary matter. Physicists have not observed such a scalar field yet but presume due to theoretical reasons that such type of field existed with very high energy density that caused inflation epoch after the big bang instance[13].

The universe after inflation was very close to homogeneous, flat, and it was devoid of all particles. Since inflation reduced the particle density to virtually zero, the particles in the universe that we observe today must all had been produced after inflation. During inflation the universe expanded exponentially and the energy density of everything else dropped to essentially zero, while the energy density of the Scalar field (called inflaton) decreases only very slowly[13]. Inflation ended when this field reaches such a low energy density that it started behaving like matter and the universe started experiencing power law expansion. So after inflation all of the energy of the universe was essentially contained in this nearly homogenous field[13]. the universe then consisted mainly of long-lived forms of energy such as electrons, protons, neutrons, and electromagnetic radiation due to quick decay of the energy in the inflaton field into other particles and fields[13].

String Theory

When string theory was extended to N-dimensional space (N can be any whole number), the nature of the sub-atomic particles became astonishingly transparent. The Kaluza-Klein theory could not determine the correct value of N[2]. The super-gravity theory, the advanced version of Kaluza-Klein theory entered into the scene but still difficulties could not be completely wiped out. Ultimately the superstring theory was proposed by the two physicists Michael Green and John Schwarz that predicted the precise value of N as ten and the conception of ten-dimensional space and time appeared to the physicists as a very powerful tool to work for exploring the unified theory for unification of the four fundamental forces of nature. The physicists thought that Kaluza-Klein's curled up dimension provided the clue for existence of additional dimensions. Considering the curled-up space as sphere, we get another three dimensions giving a total of six dimensions. But still there is requirement of more dimensions. Long before inception of string theory, Calabi and Yau, the two mathematicians described six dimensional geometrical shapes Fig. B.1. If the Kaluza-Klein curled-up space is replaced by Calabi—Yau shapes, we can get ten dimensions, three spatial, six Calabi-Yau shapes and one time[12].

Since the universe remained very unstable in ten dimensions, it was broken into two pieces giving rise to formation of our four-dimensional universe being separated from the ten-dimensional universe. According to superstring theory, the big bang is the by-product of the much more violent cracking of the ten dimensional universe into two pieces. It also indicates that after the split, the remaining ten-dimensional universe has shrunk to such an exceedingly small size i.e. about 100 billion billion times smaller than the nucleus of the atom, that it cannot be detected with available tools[3]. Some string theories even need as many as twenty six dimensions to explain particles and particle interactions. A great expectation has been created by the superstring theory that it will be able unify the four fundamental forces of nature which will be seen as the different manifestations of a single unifying force.

Fig. B.1 Calabi-Yau six dimensional shape

The superstring theory has used the idea of dark matter, a new form of matter which is supposed to have been created at the time of cracking of the ten dimensional universe due to fission. This dark matter has weight like all other matter but there is strong evidence that dark matter might not be made up of Fermions. We do not know what is dark matter. It may be made of neutrinos or neutralinos, one of the theoretical supersymmetric particle. Dark matter cannot be detected even with most sensitive instruments[3].

Mathematical calculations show that if the universe has sufficient matter, the gravitational attraction of the galaxies will slow down the expansion or even reverse the expansion to an ultimate process of collapse. The astronomers do not find adequate quantity of matter in the universe which can reverse the expansion. On the other hand, the data received from red shifts and luminosities of stars suggests possibility of collapsing universe[3]. The role of dark matter has been considered quite relevant in explaining such a situation. Thus superstring theory attempts to successfully answer the two basic questions arising out of big bang theory regarding situation before the big bang and initial condition at the time of big bang.

The two major achievements of the twentieth century physics are inception of quantum mechanics and theory of relativity to explain the mysteries of nature. The marriage between these two ideas was called Quantum field theory. In the process of establishing the theory, the physicists brought the idea of antimatter resulting in doubling of the number of elementary particles. The quantum field theory, however, could not contrive smoothly because of mathematical inconsistencies. The Idea of supersymmetry solved the problem and the number of particle got double once more. Now what is supersymmetry?

Supersymmetry

In an effort to unify the gravity with the other three fundamental forces, the physicists brought an idea that every elementary matter particle have a massive 'shadow' force carrier particle and every force carrier particle have a massive 'shadow' elementary matter particle. The relationship between the matter particle and the force carrier particles is called supersymmetry (Fig. B.2). For example: for every type of quark, there may be a type of particle called 'squark'. These supersymmetric particles have not yet been detected but scientists are exploring the possibility of existence of such particles[21].

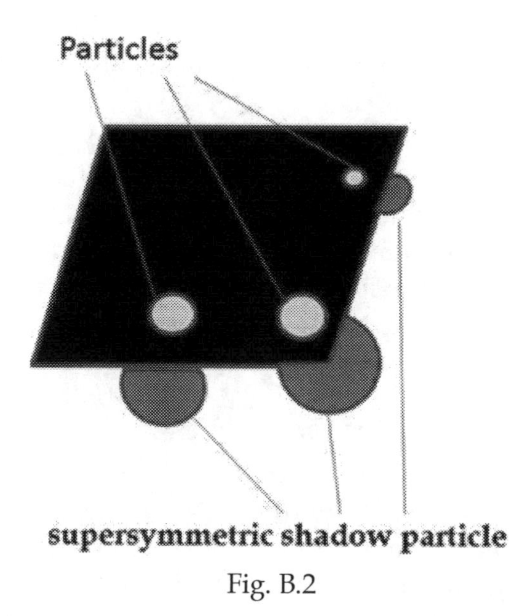

Fig. B.2

Multiple string theories

Through the present day's particle accelerator, we can probe up to distance of the order of 10^{-13} cm. At his range of distance scale, the force of gravity cannot be probed and to do this the accelerator should be capable of exploring a distance scale of the order of 10^{-33}cm. We are far away to have such an accelerator. So the unification of the four fundamental forces of nature cannot be experimentally verified within the four walls of laboratory. However, in macro scale, there are lots of cosmological phenomena that involves very high density of energy or densely packed elementary particles or matter which can guide us in forming conception of a unified field theory but definitely adequate precision will not be available.

A unified theory must satisfy all the demands of quantum field theory of standard model and general theory of relativity accurately in the respective physical circumstances where each of them describe the related physics precisely. The unified field theory should also address accurately the physical circumstances where both of them are applicable. There are various attempts to predict such a unified theory but super string theory is the only one that has been able to come very close to fulfilment of all the criteria.

String theory replaces the principles of point like particles which are the basic ingredients of quantum field theory and general relativity and states that the elementary excitations of our universe are not particles and the fundamental building block of nature are small vibrating strings. These strings are made of little lines of energy. When one tries to divide them, they form new little stretches of energy. There is no elementary constituents smaller than these line of energy. These strings are 100 billion billion times smaller than the size of a proton[3]. They are so tiny that even our latest particle accelerator cannot sense them. We see them as point particle because of the limit of maximum resolution that can be achieved. The quantum mechanical incorporates both open and closed strings (Fig. B.3). Such mathematical models incorporates lots of excitations that look like point like particles. One of such excitations is the massless vector particles. They are the key messengers in describing the standard model. Open strings can merge into open loops forming closed strings. These open and closed strings are the basic ingredients of the standard model. Quantum mechanical theory of closed rings also incorporates a lot of massless particles and one of them is a massless spin 2 particle. This is miraculously identified in the formulations that is required to form the theory of gravity.

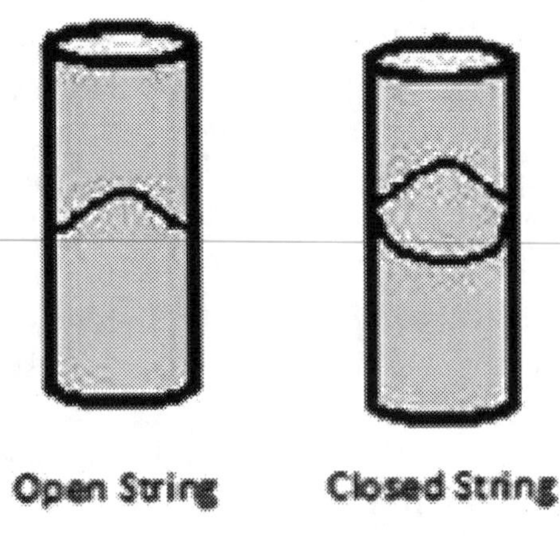

Fig. B.3

As per string theory, the fundamental forces and various particles as we see in nature are nothing but the pattern of oscillation that the string is undergoing. According to Super String Theory, the properties of the different particles depend on the resonant pattern of the vibration and can be compared to different notes that an ordinary vibrating violin string can play—electrons, photons, quarks. All matters and even the force particles are associated with the pattern of the string vibration. The string oscillations are the possible indications of the unification of the forces and the harmonious union of general relativity and quantum mechanics is a major success of the string theory[3][12].

If we consider that strings can only vibrate north and south, east and west, up and down, we shall not get enough variations in vibrations to account for existence of all the particles and forces. At least six more spatial dimensions are required for the equations of String Theory.

Initially, the string theory could explain the particles like photons, gluons, gravitons which are Bosons and hence was called Bosonic string theory. All the vibrational pattern of the Bosonic string had integer spin.[14]. Bosonic string theory had two problems. It did not incorporate the Fermions and hence there was no Fermionic pattern having half integer spin consisting of particles like electrons, quarks[[12]. The patterns of Bosonic String theory was considered for a particle Tachyon having negative mass. While exploring possibilities of existence of various particles in the universe, the physicists imagined the existence of Tachyon in addition to particles having positive masses. But ultimately the theory could not establish itself. Similarly, at the initial stage, the inclusion of such a particle in their theory of strings developed a set back. In 1971, Ramond took up the challenge to include Fermions in the string theory and later on Schwarz and Neveu presented a new version of string theory incorporating super-symmetry which shows that the Bosonic and Fermionic patterns of vibrations are coming in pairs. For each Bosonic pattern of vibration there is fermionic pattern of vibration and vice versa. The super-symmetry gave birth to super string theory which was able to overcome the problem due to inclusion of

Tachyon in the Bosnic string theory, explain the fundamental forces of nature and accounts for the visible matter particles constituting the universe.

One of the most realistic arguments in support of existence of a final theory of nature in some particular form is that the theory couldn't be otherwise[12]

During the period from 1970 to 1984, depending on the general properties of strings, five theories were discovered[23]. There are three different string theories comprising of closed strings and open strings. Mixing up the best features of Bosonic string theory and superstring theory, two other consistent theories were proposed which are called Heterotic string theories. Though the five different theories have something in common i.e. they have the vibrational patterns which determine the possible masses and force charges, all of them require ten dimensional space-time, their curled up dimensions must be in one of the Calabi-Yau shapes, still they are five different self-consistent superstring theories based on different symmetries, not resembling with one another. In late 1980s, physicists realized that though the string theory had been able to provide more or less convincing picture of the universe, it was not somehow acceptable to them that five different theories should exist and they thought that the multiple self consisting theories were quite irrelevant for unification of gravity and quantum theory[14].

Being puzzled with five different string theories in ten dimensions, Edward Witten of Princeton and Paul Townsend of Cambridge quite unexpectedly discovered in 1995 that the 'true home' of the superstring theory is actually the 11th dimension and the strings are coexisting with a strange collection of membranes[3]. Witten called it M-theory which created the second superstring revolution. M means mother or mysteries or magic and so far it could be explored only a very little. The results of M-theory is very encouraging. The problems which were thought to be difficult to be solved mathematically, have been seen significantly easier. The laws of physics gets simplified at higher dimensions. In fact, when viewed from the platform of the eleventh dimension, all five string theories appeared to be connected with one another as if they are different limiting cases of a single theory or each a special case of some more fundamental and unique theory. It is like the advantage gained by the ancient Romans, adding extra dimension to the warfare by seizing the mountain top, the height as additional dimension, to have better understandings of the status of the war in fronts and coordination with various fronts of the battlefields[3].

String coupling constant

The basic interactions between the vibrating strings involve splitting apart and joining together of string loops(Fig. B.4 a. two strings on a collision course b. string motion emphasized c. 'time lapse' photograph on world sheet). The two strings slam together at some point and then merge together into a single loop. This loop after travelling a bit creates a virtual string pair (string-antistring pair) due to hyperactive quantum fluctuations. The virtual string pair subsequently annihilates and produces a single string. Finally it gives up energy by dissociating into a pair of string[12].

There is a number 'g' that determines the likelihood of the quantum fluctuation that causes a single string to split into two strings momentarily yielding a virtual pair. This number is known as string coupling constant. The size of the string coupling constant indicates how strongly the strings are coupled to one another. The smaller is the coupling constant, the less is the probability of formation of virtual string pair.

M-theory is expected to explain origin of strings and solve many long standing puzzling mysteries about string theory. It has been seen that superstring theories are related by duality transformations known as T duality and S duality.

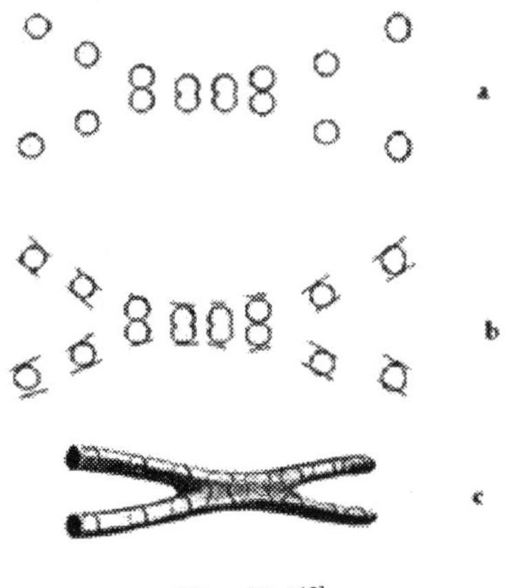

Fig. B.4[12]

Duality

In its oldest form, duality is a symmetry between electric and magnetic fields which hold in vacuum but invalid in nature as we see electrons but magnetic monopoles are not visible[23]. In a very crude way the duality can be described as the possibility of two descriptions for the same physical conditions. In the following example (Fig. B.5), let the location of point in a X, Y, Z coordinate system is described by (x, y, z). If we choose another coordinate system X′, Y′, Z′, the location of the same point will be (x′, y′z′). The

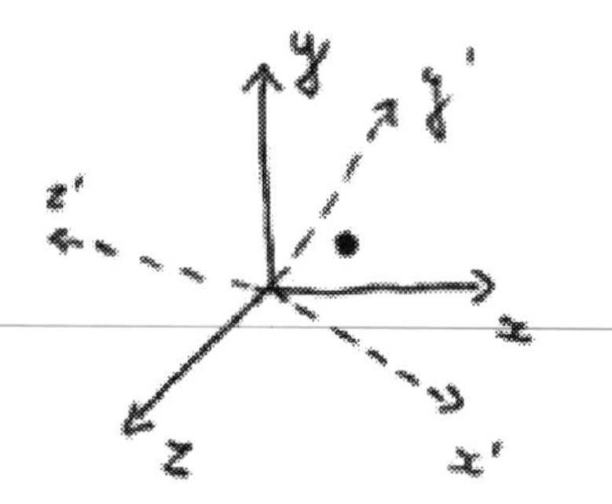

Fig.B.5
(duality in location of the point)

position of the point in space is same but there are two descriptions of its location.

Thus there may be situation that involves two seemingly distinct physical theories which indicate different descriptions of the same physics. Such theories will be called dual and then one theory can be mapped to other making it clear that they are equivalent physical theories but having different descriptions.

We know that the string theory requires extra six spatial dimensions which are tightly curled up in space. We can think that these dimensions may be arbitrarily big or arbitrarily small with different physics holding in each of the cases. When the dimension becomes very small i.e. of the order of Planck length, something peculiar happens. To simplify the case, let us consider the case of a string theory of closed strings which is existing in five dimensional space-time, the extra dimension of which is a circle of size of Planck length curled up in space. Now let us start contracting the dimension and measure the resulting physics. The readings will vary depending on the size of the dimension. If we repeat the experiment by expanding the size instead of contracting, the observed physics will appear exactly the same as for the case when size was contracted. Therefore there is duality between these two scenarios.

Let us have two string theories, theory A and theory B. For the theory A, the circle be of radius R and therefore the length of the fifth dimension is $2\pi R$. The other string theory is theory B which is precisely like theory A with the exception that the fifth dimension on which it lives is a circle of radius $(1/RT)$, T being the parameter equal to the tension of the string. Let us now consider that strings are extremely small and look like a point on the circle and behave like a point particle. The point particle when considered quantum mechanically, behaves like wave and to live on the circle, the wave length of that wave should be fraction of the perimeter of the circle. As per De Broglie's conception, the momentum should be equal to multiple of the inverse of the radius of the circle. So for theory A, there are string states with momentum or mass n/R where n is a integer and for theory B, the string states will be of momentum or mass mRT where m is an integer. As we have considered closed strings wound around the circle, they have to wind around the circle an integer number of times. Since the strings are having some tension, they take some energy to stretch over the circle. Now the mass of these strings are proportional to the tension of the strings and number of times they wind around the circle. Hence the mass of the strings are pRT for theory A, pR being the number of windings and q*(1/RT)*T=q/R for theory B, p and q are generic integers.

Thus it appears that there can be one-to-one mapping between theory A and theory B as in Fig.10. Indeed, the string states in theory A with mass n/R can be mapped onto the stretching strings of theory B with mass q/R (when we map the state with momentum n to the state with winding q times). Similarly, for theory A, the winding states can be mapped to the state with momentum mRT of theory B. The theory A and theory B have precisely the same physical situations and therefore are dual(Fig. B.6). Let us now consider string theory A with circular dimension of radius R less than √(1/T) i.e.

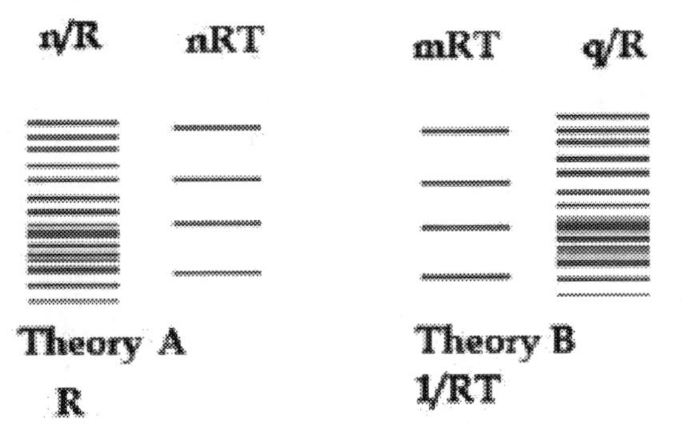

Fig.B.6

square root of inverse of string tension and string theory B with circular dimension of radius 1/RT bigger than √(1/T). Therefore to study a smallest distance in the universe, we need not study theory A as the radius gets smaller and smaller. We can study the theory B with bigger radius and still get the idea of the universe as the two theories have duality and they are equivalent.

The above instance of the dualities between large and compact spaces gives the idea of dualities between straight and tilted spaces, between very complicated spaces to other very complicated spaces. The string theory on space A depends on the precise properties of space A and so for string theory B on space B. If there is stringy T duality between those two theories, some properties of space A can be mapped on to space B. Thus stringy T-duality allows us to study properties of a complex space A, by studying other, perhaps more convenient properties of space B. Utilizing this idea that universes of very small size look equivalent to universes of larger sizes, we may be able to get a better perception of physics at the big bang and get a less singular and more satisfying picture of what went on at extremely early times by studying consequences of a theory of quantum gravity.

A more fundamental duality is referred as S-duality. All quantum field theories specify coupling constants which decide the strength of interactions between the particles. For string theories also same observations apply. The value of these coupling constants vastly affect the strings behaviour as indicated earlier. The five string theories were proposed after the first superstring revolution and it was thought that they were completely separate. With discovery of various S-duality, different distinct types of string theories were paired up. Suppose string theories A and B have adjustable string coupling constants. If A has a large coupling constant and B has small coupling constant, then they predict precisely the same physics. These duality relationships between string theories have initiated a radical change in the perception of string theory and have provoked the expectation that all five superstring theories are special limits of a more fundamental theory.

Einstein's famous example of the lion and its tail gives us a picture of the duality in the universe. When we see only the tail of the lion, we can be sure that there is a lion connected to the other end. Similarly when we see the laws of nature around us including the fundamental forces, there must be the lion which is here the unified field theory behind the scene that is apparently not visible or realisable[3].

The story of lion and the blind wise men has been utilized for projecting the justification behind the existence of different valid and workable string theories. One day three blind wise men happened to come across a lion by accident. As the lion was running away, one of the blind wise men chased and desperately grabbed the tail of the lion and shouted, "I have got the string". Another one caught hold of the ear of the lion and exclaimed," Oh! It is a membrane". By this time the third one also reached the spot and tried to hold the lion by grabbing one of its leg and yelled with joy," Both of you must be wrong, it is actually the three-brane!". Now if the incident is examined, we shall find that all those three were correct in view of their stand point as each of them was probing a particular portion of the body of the lion. They described correctly standing on their specific domain. But again the reality is the lion itself giving rise to mystery[3][12].

With the new symmetry, duality and the new M-theory, the scientists are with big hope that the closing version of the string theory will be evolved out of these new theories and the mystery of the lion will be gradually unfolded to provide the final theory.

It is true that the superstring theory describes a comprehensive formulation of the theory of universe, but its experimental verification is beyond the capability of present-day technologies. Theory predicts that all the forces can be unified at Planck energy, i.e. 10^{19} billion electron volts and this is about one quadrillion times higher than the energies presently available in our accelerators[3]. This is very much disappointing that we cannot test the theory which has the tremendous potentiality to drive the future of physics with our current generation of equipment or with machines of next few generations ahead technology in conceivable future.

Black Hole : According to Einstein 's Theory of General Relativity, light is bent by strong gravitational pull which has been established by the observation during solar eclipse that starlight is bent when grazing the Sun's surface. A black hole is a region of space where the gravitational pull is so strong that no matter can leave the black hole. When light passes by a black hole, light cannot escape. Thus we cannot see a black hole but its existence can be detected through other observations.

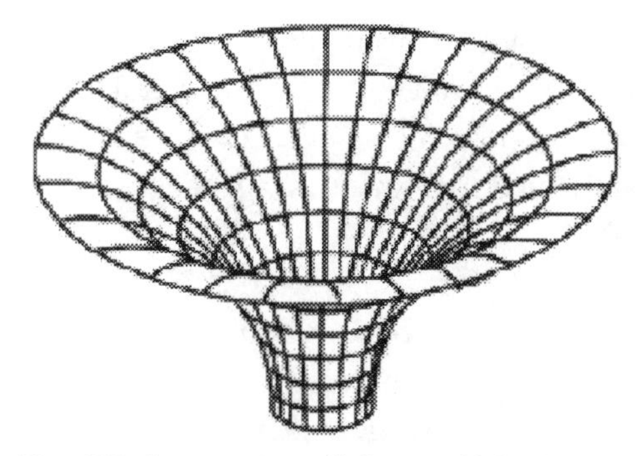

Fig. B7 Space-time fabric of black hole

Black hole emits x-ray radiation, looses mass and gets smaller and smaller until it disappears, or evaporates. Most black holes are formed due to death of stars, larger than the sun, Those stars run out of fuel and cannot sustain their nuclear reactions. The centrifugal force generated within the star due to thermonuclear reaction, gets overcome by the force of its own gravity pulling it inward. Whatever materials remain, collapse on itself. Now, how much mass remains when the star dies, decides what will be its new incarnation. When the size of the star at that time becomes approximately same as that of sun, it becomes 'white dwarf' which continues to glow with left over heat. If the size is about three times of that of sun, it becomes 'neutron star'. When the size of the dead star is more than three times of that of sun, it becomes black hole. In case of black hole, the star ultimately gets so much gravity and becomes so much compact that it consumes itself, all its mass squeezed and nothing left there but a single point in the fabric of space-time remains where both space and time stop. This point is called the Singularity. This is the core of black hole.

Astronomers presume that black holes are present in several galaxies at their centres; They also believe that the Milky Way, our own galaxy also has a large black hole. The influence of a black hole is felt beyond a

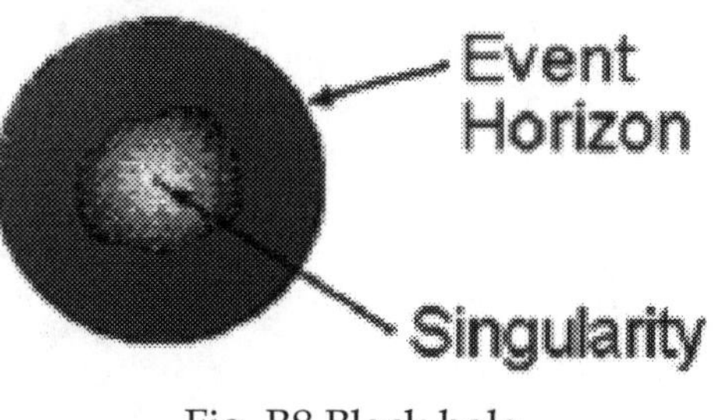

Fig. B8 Black hole

"Event horizon" and once that horizon i.e. the critical distance from the black hole is crossed, matter or light cannot come back out. Near the horizon, matter in the vicinity is accelerated and sucked into the black hole. Accelerating matter emits great amounts of radiation.

The existence of Quasar or quasi-stellar radio source is detected at the edge of the observable universe. Quasars emit vast amounts of energy in the radio spectrum. It is guessed that this energy comes from matter falling into a large black hole. Black holes don't emit electromagnetic radiation, such as a star does, since light cannot escape due to the large gravitational attraction.

Electromagnetic waves: Electromagnetic waves are formed when an electric field couples with a magnetic field (Fig. B10). The magnetic and electric fields of an electromagnetic wave are perpendicular to each other and to the direction of the wave. Radio waves, television waves, microwaves and light wave are all different types of electromagnetic waves. They differ from each other in wavelength and constitute the electromagnetic spectrum. Wavelength is the distance between one wave crest to the next (Fig. B9).

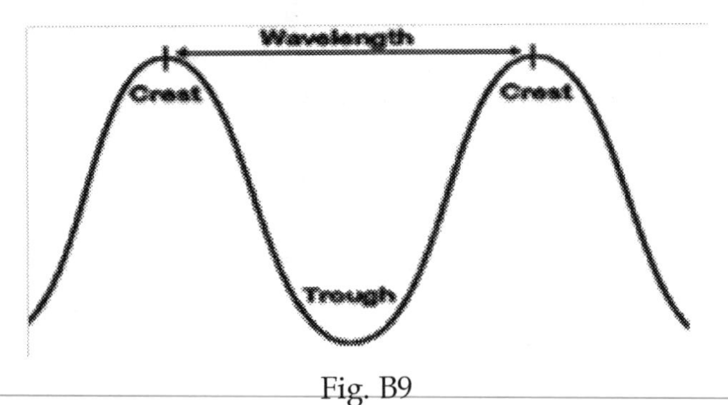

Fig. B9

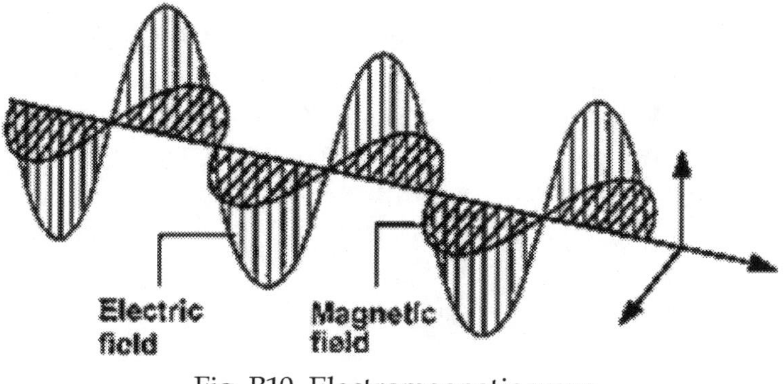

Fig. B10 Electromagnetic wave

Waves in the electromagnetic spectrum vary in size from very long radio waves of wavelength as long as the height of buildings, to gamma-rays having wavelength smaller than the size of the nucleus of an atom (Fig. B11).

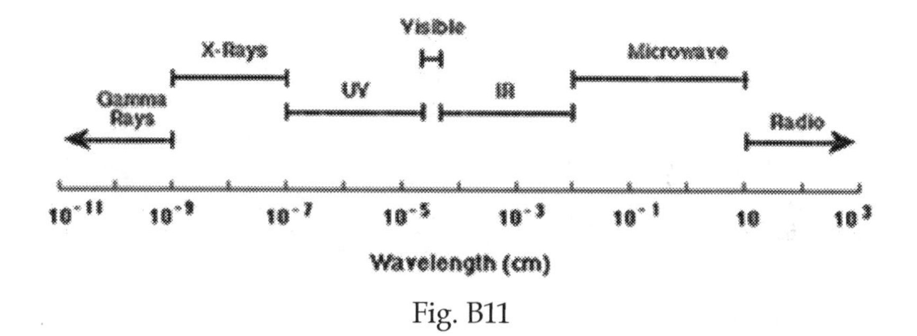

Fig. B11

Electromagnetic radiation takes place when an atom is excited by some form of energy, say, electrical energy. Being excited the electrons in the atom absorbs energy and move to the orbitals of higher energy levels. But the electron does not stay in the higher energy level and its natural tendency is to fall back to lower energy level and while doing so, it radiates the extra energy it carried to jump to higher energy level. This energy comes out as electromagnetic radiation. Depending on the type of atom and the volume of energy, the release of energy forms different kind of EM waves like heat, light, microwave, radio-wave etc.

Power law expansion of the universe: Time for doubling the distance is considered proportional to one over the square root of the energy density and this is the relation between expansion and energy density. If there are 1000 galaxies in some region of space and all distances get doubled then the volume of space occupied by those galaxies will increase eight times. Since the galaxies have the total mass same as before, their density will decrease by eight times.

If T is the doubling time when D is the energy density,

$$T = k * 1/\sqrt{D} \quad \text{where k is a constant}$$

After T, the distance between galaxies in some region of space gets doubled, Volume of that space increases 8 times and hence energy density will decrease by 8 times and becomes, say D′. So

$$D' = D/8$$

With this energy density D′, the doubling time T′ will be

$$T' = k * 1/\sqrt{D'}$$

$$\text{Therefore, } T' = T * \sqrt{(D/D')} = T * \sqrt{(8D'/D')} = T * \sqrt{8}$$

Thus if the mass of galaxies were the only form of energy in the universe then every time distances get doubled due to expansion, the doubling time would increase by a factor of the square root of eight. So if energy of the universe is composed of mass only it will experience power law expansion[13].

Anisotropy: It is the property of a material due to which the material behaves differently in different directions. The properties, physical, chemical or both of anisotropic material varies with the directions in which it is measured. A common example of such a solid anisotropic material is graphite. Example of anisotropic liquid is liquid crystal which is having structural ordering of its molecules. Similarly Plasma is the example of magnetic anisotropy. Seismic anisotropy is the variation of speed of seismic waves with direction. The electrical anisotropy can be caused due to geological formation of number of layers of materials due to sedimentation which provides different electrical conductivity in different directions. The uneven distribution of temperature in cosmic microwave background radiation is called cosmic anisotropy.

Quantum uncertainty: This is most popularly known as Heisenberg's uncertainty principle which says that position and momentum of a single particle cannot be simultaneously measured with equal precisions. The more precision is attempted to achieve in measurement of one parameter, the more accuracy we have to sacrifice in measuring the other parameter.

In agreement with the uncertainty principle, physicists are of the opinion that particle-antiparticle pairs or virtual particles, also known as quantum vacuum fluctuations, exist in empty space for a very brief period of time. It is supported by uncertainty principle that, even there is not sufficient energy to create them, the particles can come into existence for a short time. It is supposed to have been created from uncertainties in energy. These particles borrow energy for their creation and after a short time, return the energy back and vanish. For this reason, they are called virtual particles for not having permanent existence. These virtual particles, though we cannot see, leave detectable traces of their activities in empty space which has been experimentally verified with precision. As per modern physics, 'Nothing' does not exist even in perfect vacuum where the pair of virtual particles are constantly being created and destroyed.

With the new conception of vacuum and mutability of matter, it is suggested that the universe is created out of 'nothingness' and through the huge quantum fluctuations which we call 'big bang'. It is presumed that there are about 10^{85} particle-antiparticle pairs or virtual particles in the observable universe. The particles might have been created from the energy contained in virtual pairs. It is also presumed that the total energy of the universe is zero. The matter of the universe constitutes positive energy and the gravitational field provided negative energy. In universe with uniform space, the negative energy of the gravitational force neutralises the positive energy due to matter and results in total energy of the universe as zero.

We know that the gravitational field is nothing but space warp or curved space. The energy imbibed in the space warp can be converted into matter and antimatter. This happens near a black hole and probably a potential source of particles at the time of big bang. The matter appeared spontaneously out of empty space when entire universe was in state of zero energy, the matter energy being offset by gravitational energy.

APPENDIX-C

Electron orbitals in an atom

The orbital is sphere if angular quantum number *l*=0 and there can be only one way in which the sphere can be placedional energ in space. The shape is polar when *l*=1 and cloverleaf shape when *l*=2. The shape becomes complex as the value of *l* increases.

The orbital having polar or cloverleaf shape can be placed in different directions. To describe such orientation of the orbital in space, m is used where m is called magnetic quantum number because the presence of different orientation of the orbital was detected first in presence of magnetic field. All the three quantum numbers are integer i.e. 0,1,2,3, ...

The principal quantum number cannot be zero. The angular quantum number *l* can be any number between 0 and n-1. The magnetic quantum number m can be any integer between +*l* and –*l*. If n=2, *l*= 0 or 1 and m is -1,0,+1[68].

Orbitals and shell /subshell

Orbitals having the same value of principal quantum number (i.e. n=1,2,3 etc) form Shells with different value of *l* (0,1,2 etc.) which are called as shell s(where *l*=0), p(where *l*=1), d(where *l*=2), f (where *l*=3) etc.

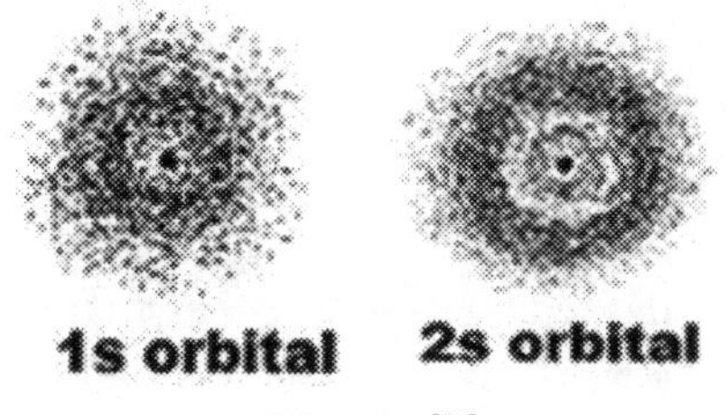

Fig. C.1[72]

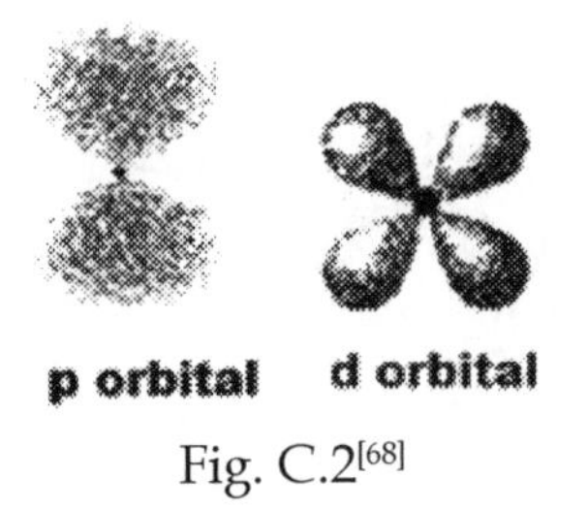

Fig. C.2[68]

The allowed combination of the quantum numbers are as shown in the table at Fig C.5. The structure of s, p and d orbitals are shown in Fig.C..2

Orbital 1s—The orbital with value of n=1, will have only one shell called 1s which is characterised by (n=1, l=0 and m=0) as shown in Fig. C.1.

Orbital 2s and 2p—Orbitals having value of n=2, will have clusters of four orbitals with two shells, s and p, where shell s is divided into two orbitals (1s and 2s shown in Fig. C.1) and 2p is divided into three subshells and hence have three orbitals as shown in Fig. C.3.

One Shell of Orbital 2s - (n=2, l=0, m=0)
Three Subshells of orbital 2p (Fig.14) are:

$2p_x$ - (n=2, l=1, m=-1)
$2p_y$ - (n=2, l=1, m=0)
$2p_z$ - (n=2, l=1, m=1)

We find that 2p sub shell has three orbitals $2p_{x,}$ $2p_y$ and$2p_z$ pointing in three direction, x, y and z coordinate axes as shown in fig C.4[72].

Orbital 3s,3p and 3d—Similarly for n=3, there are nine orbitals. One orbital in 3s shell (n=3, l=0, m=0), three orbitals in 3p subshell(n=3, l=1, m=-1; n=3, l=1, m=0; n=3, l=1, m=1), five orbitals in 3d subshell (n=3, l=2, m=-2; n=3, l=2, m=-; n=3, l=2, m=0; n=3, l=2, m=1;n=3, l=2, m=2)

The number of orbitals in a shell is square of the principal quantum number 'n' i.e. number of orbitals in shell n=1 is 12=1; for n=2, it is 22=4 and so on.

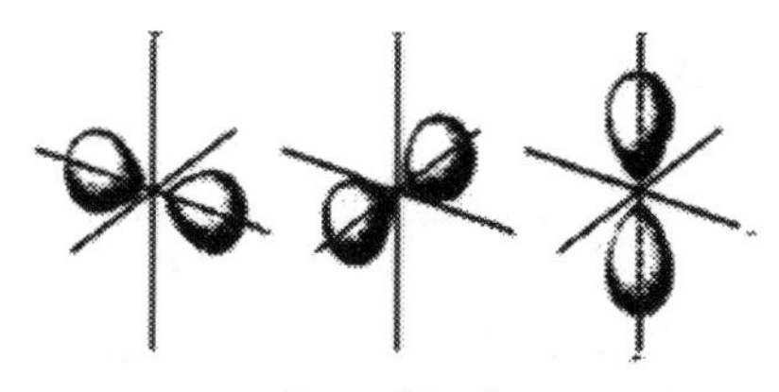

2p orbitals
Fig. C.3[68]

Number of orbitals in a subshell is $2l+1$. So for s subshell, number of orbitals is 1. For p subshell, number of orbitals is 2x1+1=3(here l=1). For d subshell, number of orbitals is 2x2+1=5(here l=2) and so on.

For 3d subshell, the orbital structure is shown in Fig.15. In three axes plane, three orbitals are in xy, xz, yz planes. The fourth orbital is along z axis and the fifth orbital lies along x and y axes.

The orbital structure will go on like the above principle as n increases.

Emission of radiation occurs when an electron dropped to an orbit of lesser energy, the lost energy being carried away by a Planck-Einstein photon. The reverse case happens when an incident photon sufficiently energise an electron to jump to higher energy orbit.

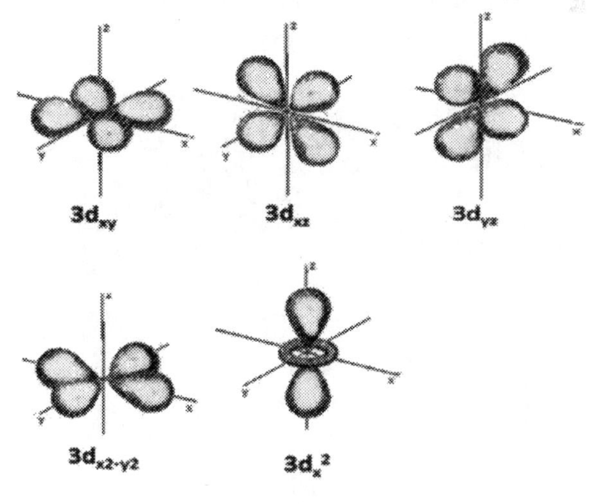

3d orbitals
Fig. C.4 [68]

Pauli's exclusion principle and Spin quantum number

The Austrian physicist Pauli proposed in 1925 that apart from the three quantum numbers, n, l and m of Bohr-Sommerfeld model which together characterised the size, shape and spatial orientation of the electron's orbit in the atom, the electron also possessed additional two valued quantum numbers which was referred as spin quantum number 's' having half integer value, +1/2 or -1/2. With this, the effect of spectra of the atom could be explained. Pauli then stated that no two electrons in the atom could have the same set of four quantum numbers or could be in same quantum state. In the same year, the two physicists, Bothe and Geiger demonstrated that the energy and mass are conserved in atomic processes.

The energy level of orbital goes higher with value of n. It means that for lower value of n i.e. orbital closer to the nucleus have lower energy. Electron, before filling up higher energy levels occupy lower ones. For orbitals of exactly equal energy, the electron has a choice and it fills up orbitals singly as far as possible keeping the repulsions between electrons minimum so that atom is stable.

As seen in the Fig. C.6 below, the s orbital always has a slightly lower energy than the p orbitals at the same energy level, hence before filling the corresponding p orbitals electron always fills up s orbital[72].

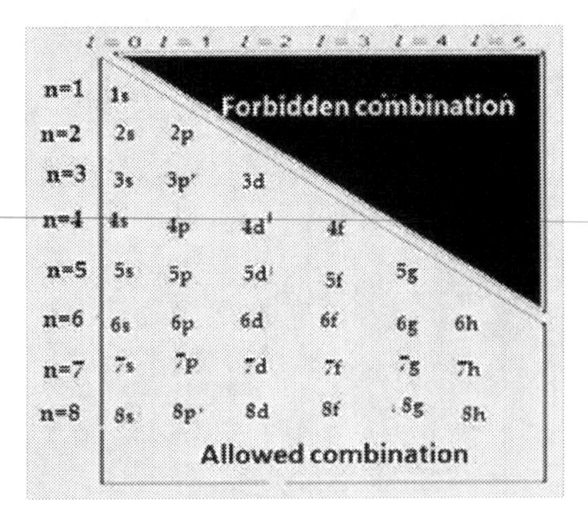

Fig. C.5

As seen in the figure, since the energy levels of 3d orbitals are slightly higher than that of 4s orbital, the electrons first occupy 4s orbital and then 3d orbitals.

As per Bohr's model of a hydrogen atom, the frequency of light radiated by an electron moving from an orbit n_i to an orbit n_f corresponds to the energy level difference between n_i and n_f.

The number of protons in the nucleus of the atom of an element is atomic number of that element. It is an unique number and the

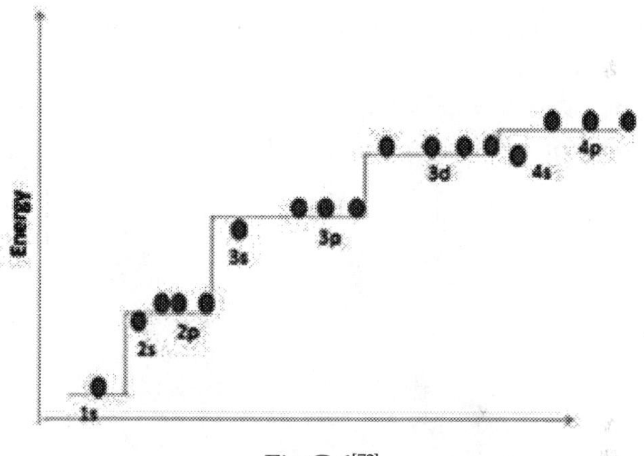

Fig.C.6[72]

atomic number of an element does not change, implying that the number of protons in the nucleus of every atom in a normal element is always the same.

In 1926, Italian-born physicist Fermi and British physicist Dirac provided mathematical model of Pauli's exclusion principle and equations that described electron's behaviour.

Half life : Half life of a radioactive material is the time by which any amount of the sample of that material will disintegrate to half of that amount. If T_{half} is the half life and A_0 the initial amount i.e. at time t=0, then after time t, the amount A will become:

$$A= A_0 2^{-t/T_{half}}$$

Bell curve distribution: It is also called Gaussian distribution or normal distribution. 'Bell curve' is the shape of the curve which is created when a line is plotted using the data points for distribution that satisfies the criteria of 'normal distribution'(see Fig. C.7). The centre of the curve contains the greatest number of a value and therefore would be the highest point on the arc of the line. This point is called the mean, and in simple terms it is the highest number of occurrences of an element. (statistical terms, the mode). It is interesting to note that a normal distribution is the curve is concentrated in the centre and decreases on either side.

Quarks: The first generation quarks are the lightest quarks, Up quarks and Down quarks designated as u and d and build protons and neutrons in an atom. The up quark has an electric charge of +2/3, and the down quark has a charge of -1/3. The second generation

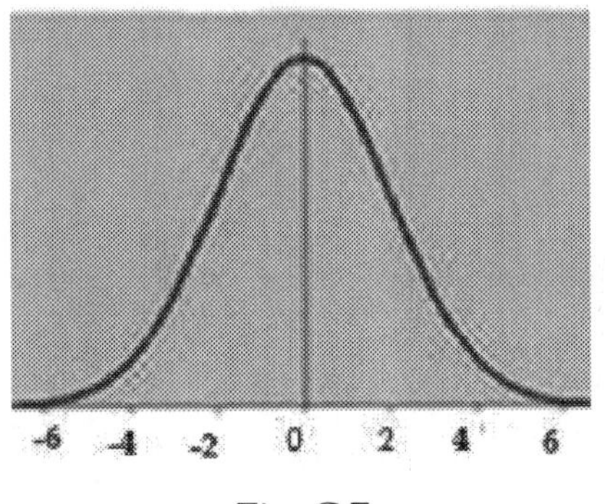

Fig. C.7

quarks are Charm quark and Strange quarks designated as c and s. The charm quark is named so on a whim and Strange quark has got its name from the strangely long life time of the composite particle K, constituted by this quark. These quarks have greater mass than those categorised in the first generation. The electric charge of charm quark is +2/3, and the strange quark has a charge of -1/3. The heaviest quarks are the third generation quarks and called as Top and Bottom quarks designated as t and b[88][89].

The top quark has an electric charge of +2/3, and the bottom quark has a charge of -1/3. u, t and c quarks have identical behaviour and hence called up-type quarks. The d, s and b quark are called down-type quarks because they share the same type of electric charge (Fig. C.8).

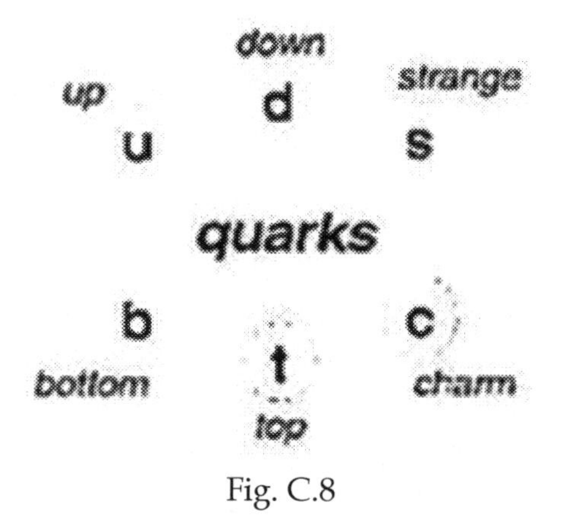

Fig. C.8

Gluons: Six of the eight gluons have two colours, one has four, and another is having six colours. Apart from gluing quarks together, gluons also stick to themselves [77]. One consequence of this is that they cannot do much beyond the nucleus. Gluons carry the strong force by moving between quarks and antiquarks and changing the colour of these particles. In hadrons, Quarks and antiquarks constantly exchange gluons and as they emit or absorb gluons, the colours are changed. In case of baryons and mesons, as they are colourless, whenever a quark or antiquark changes colour other quarks and antiquarks in the particle must change colour to preserve the balance.

The quantum field theory or the mathematical model applied to describe the interaction of coloured particles like quarks through the exchange of gluons i.e. strong interaction is known as quantum chromo-dynamics (QCD)[77]. The whole sticky mess is called the strong force or the strong interaction since it results in forces in the nucleus that are stronger than the electromagnetic force.

Without the strong force, every nucleus would blow itself to fragments(repulsive force between two protons is of the order of 200 Newton!).

The particle physicists use eV as a unit of mass also. It may be questioned why we are using eV(electron volt) for representing mass, when actually it indicates the energy of a particle (1 electron volt is the kinetic energy gained by an isolated electron when accelerated by a potential difference of 1 volt). We know the famous equation $E=mc^2$ where m is the mass, c is the velocity of light and E is the energy, i.e. mass and energy are mutually convertible. So one eV unit of the mass is equivalent to one eV of energy divided by c^2. Now as per this calculation, one $GeV/c^2 = 1.783x\ 10^{-27}$ Kg where GeV is Giga electron volt i.e. thousand million electron volt or 10^9 eV. Usually the c^2 is dropped and masses quoted in GeV.

Nature of force carrier particles(Bosons)

Force Carrier Particle	Electric charge	Mass	Spin	Observation
Graviton	0	0	2	Not yet seen
Photon	0	0	1	Observed
Gluon	0	0	1	Indirectly observed
W^+	+1	80Gev	1	Observed
W^-	-1	80Gev	1	Observed
Z	0	91Gev	1	Observed
Higgs	0	>78Gev	0	?

Fig. C.9[79]

Feynman's diagram

Drawing Feynman diagrams is a popular way in visualizing and predicting the subatomic world.

Before we try to describe the concepts of the Feynman's diagram, let us discuss a little about CPT symmetry. The CPT theorem, regarded as one of the basic principles of quantum field theory which states that all interactions should be invariant under the combined application of CPT where C stands for charge conjugation,

P for parity, and T for time reversal in any order. CPT symmetry is a correct symmetry of all fundamental interactions.

Nature of fermion particles

FERMION					
LEPTON (spin- ½)			QUARK (spin-1/2)		
Particle	Electric charge	Mass Gev/C^2	Particle	Electric charge	Mass Gev/C^2
Electron neutrino	0	$<1x10^{-8}$	u up	2/3	0.003
Electron	-1	0.0005	d down	-1/3	0.006
Muon neutrino	0	0.0002	c charm	2/3	1.3
Muon	-1	0.1	s strange	-1/3	0.1
Tau neutrino	0	<0.02	t top	2/3	175
Tau	-1	1.77	b bottom	-1/3	4.3

Fig. C.10[7]

There are three different kinds of symmetries existing in the universe: charge (C), parity (P), and time reversal (T). If we charge conjugate or change matter to antimatter, reverse the parity (like a mirror reflection) and reverse the flow of time, a matter particle will be exactly same as antimatter particle (see Fig. C.11). We know that all charged particles like electrons, quarks, etc with spin 1/2 have antimatter counterparts of opposite charge and of opposite parity. Particle and antiparticle, when come together, can annihilate, disappear and release their total mass energy in some other form of energy, mostly gamma rays.

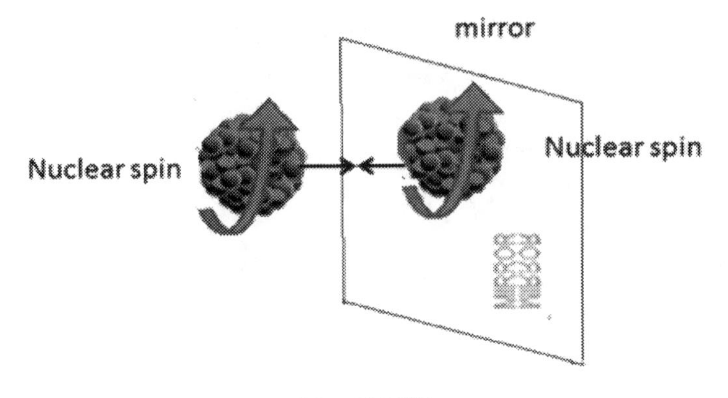

Fig. C.11[83]

The figure above clarifies this situation. The direction of the emitted electron (arrow) reverses on mirror reflection, but the direction of rotation i.e. spin (angular momentum) is not changed. Thus the nucleus in front of the mirror represents the actual directional preference, but its mirror reflection represents a directional preference which is not found in nature. For example: suppose we make a nucleus out of antimatter (antiprotons and antineutrons) then beta decay of this anti-nucleus would behave in the same way, except that the mirror image would represent the preferred direction of electron emission, while the anti-nucleus in front of the mirror would represent a directional preference not found in nature[83].

Now the questions are whether a nucleus behaves in a different way if its spatial configuration is reversed (P), if the direction of time is made to run backwards instead of forward (T), or if the matter particles of the nucleus are changed to antimatter (C).

For many years, it was assumed that elementary processes involving the electromagnetic force , strong and weak forces displayed symmetry with respect to both charge conjugation(C) and parity(P)—these two properties were always found conserved in interactions of particles s. The same appears true for a third operation also, time reversal (T), which corresponds to reversal of motion. Time invariance implies that whenever laws of physics approves a motion, the reversed motion also becomes an allowed one.

The invariance of C, P, and T were subjected to a series of examinations from the mid-1950s which caused physicists to change significantly their assumptions on CPT. The two American theoretical physicists Yang and Lee wanted to verify experimentally the parity conservation itself prompted by observation of apparent lack of the conservation of parity in the decay of charged K-mesons into two or three pi-mesons.

In 1956 Yang and Lee showed that there was no evidence supporting parity invariance in so-called weak interactions. Experiments were conducted in the following year with radioactive atoms of colbalt-60 with introduction of a magnetic field to assure that they are spinning in the same direction (Fig. C.12).

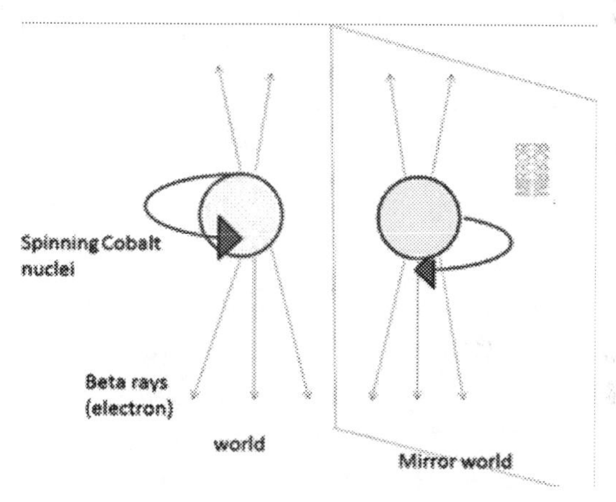

Fig. C.12[84]

It was demonstrated conclusively that parity was not conserved in particle decays, including beta decay of a nucleus, that happens through the weak force.

Such experiments also showed that charge conjugation symmetry was broken during these decay processes as well. It was also found that the weak force does not obey symmetry C. But surprisingly, the weak force appeared to obey the CP symmetry. Therefore the physical laws would be the same for a particle and it's antiparticle with opposite spin. But while conducting experiment subsequently with the decay of neutral kaons, the two physicists named Cronin and Fitch showed that the weak force caused a small CP violation[84].

The violations of these symmetries can cause nature to behave in other ways. Violation of C symmetry indicates that the laws of physics are not the same for particles and their antiparticles. P symmetry violation will imply that the laws of physics are different for particles and their mirror images (means the ones that spin in the opposite direction). The T symmetry violations indicate that if you go back in time, the laws governing the particles change[84].

The antiparticle of an electrically neutral particle may be identical to the particle, as happened in the case of the neutral pi-meson, or it may be as distinct as the case of the antineutron. Parity or space inversion occurs when the three spatial dimensions x, y, and z respectively become $-x$, $-y$, and $-z$. To be more precise, parity conservation means that left and right or up and down are indistinguishable in the sense that an atomic nucleus emits decay products up as often as down and left as often as right[85].

CP violation has important theoretical results. The physicists make an absolute distinction between matter and antimatter by observing violation of CP symmetry. This difference between matter and antimatter may have a serious consequences in cosmic world. One of the unresolved theoretical queries in physics is why the universe is made mainly of matter. There are debates on the issue but with reasonable assumptions, it can be shown that the observed imbalance or asymmetry in the matter-antimatter ratio may have been produced by the occurrence of CP violation in the first seconds after the big bang[85].

Let us now come back to Feynman's diagram. The diagram is based on the following presumptions:

1. The space-time diagram is the diagram by plotting time on horizontal axis and position on vertical axis.

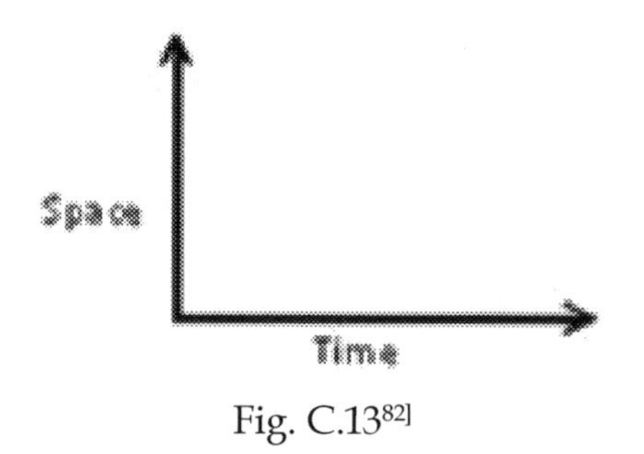

Fig. C.13[82]

2. CPT symmetry is conserved. After CPT operation the matter particle is identical to antimatter particle.

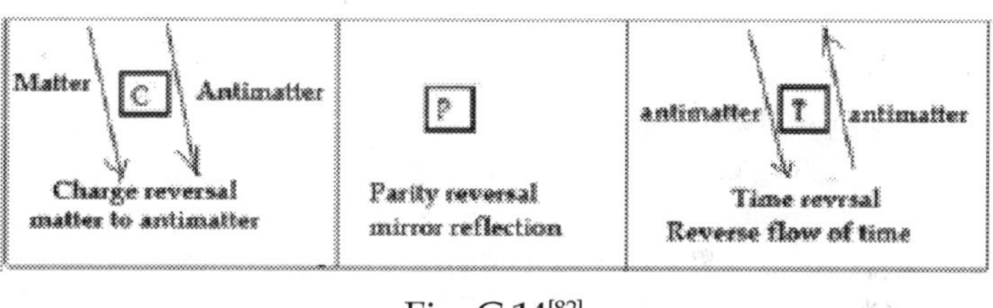

Fig. C.14[82]

3. Symbols used in Feynman's diagram:

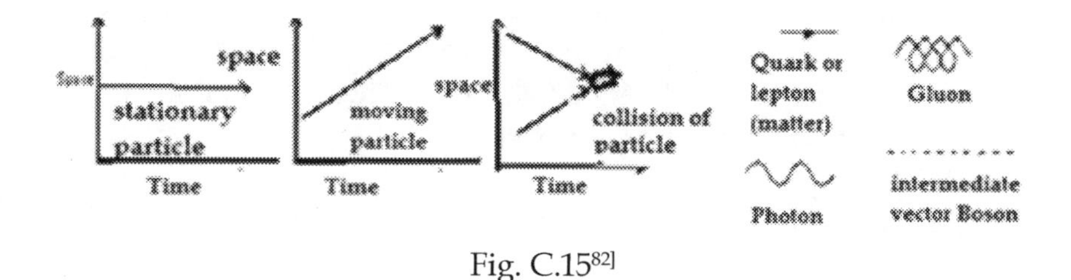

Fig. C.15[82]

4. Annihilation diagram: Electromagnetic Interaction—Matter and antimatter colliding to result annihilation to produce pure energy in the form of electromagnetic radiation photons). Then

photon creates electron-positron pair which is known as pair production.

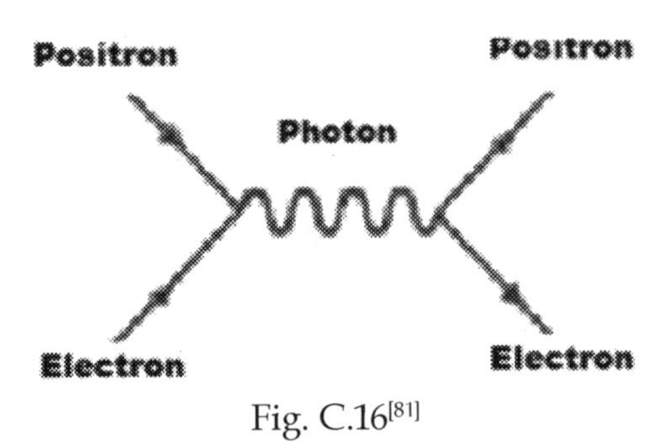

Fig. C.16[81]

5. Scattering diagram: Electromagnetic radiation—Two electrons coming towards each other and repelled by the electromagnetic force(via exchange of virtual photon)

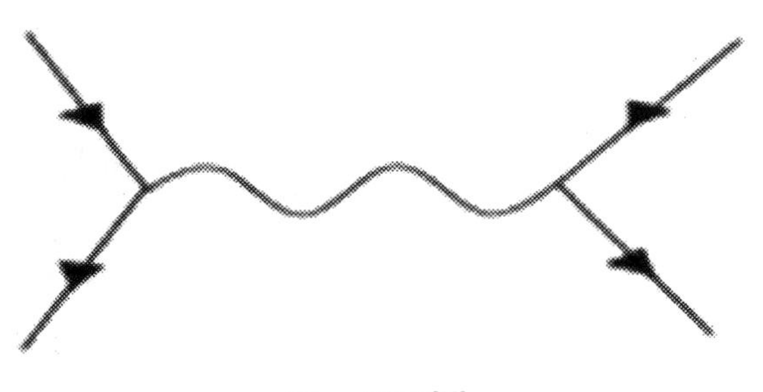

Fig. C.17[82]

One electron emits a photon and recoils; the second electron absorbs the photon and acquires its momentum. Thus the recoil of the first electron and the impact of the second electron with the photon results in the electrons driven away from each other. This is repulsion between two charged particles.

Weak Interaction

Weak interaction in beta decay can be displayed as:

$$\mathbf{n - p + e^- + \underline{\nu}}$$

The proton and neutron are not elementary particles and the process of beta decay should be analysed at the quark level. Also we know that the quark composition of a neutron is (udd) and a proton is (uud).

The neutron transforms into a proton if a down quark becomes an up quark. The emission of a W^+ or W^- changes the colour of the quark. This is called charged current reaction.

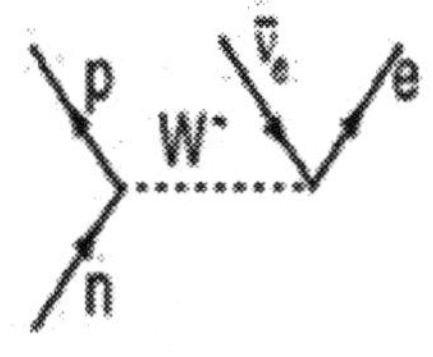

Fig. C.18[79]

d -> u + W⁻

In the neutral current reaction, Z^0 interacts with any of the quarks i.e. in six ways and all neutrinos are produced.

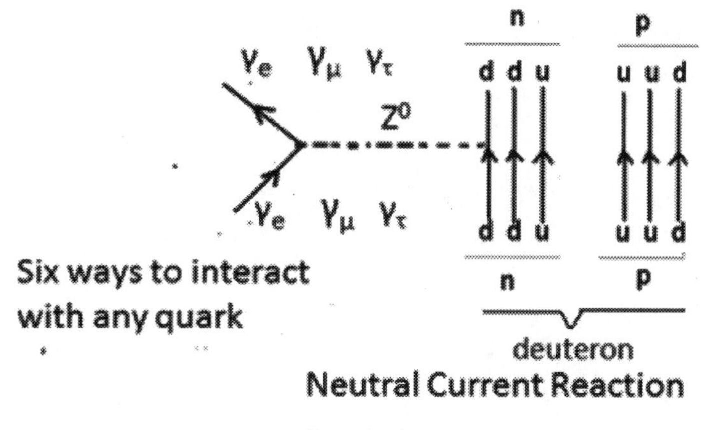

Fig. C.19

Vertices involving the Z^o particle leaves a quark colour unchanged like events with photon. The following diagrams show the basic weak events at the quark level:

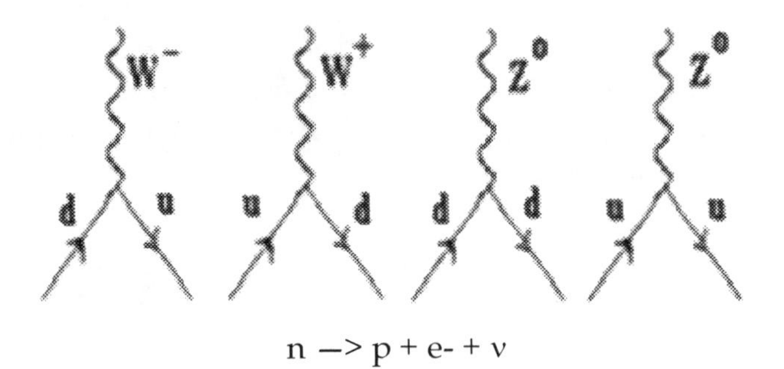

n —> p + e- + ν

Fig. C.20[81]

Strong Interaction

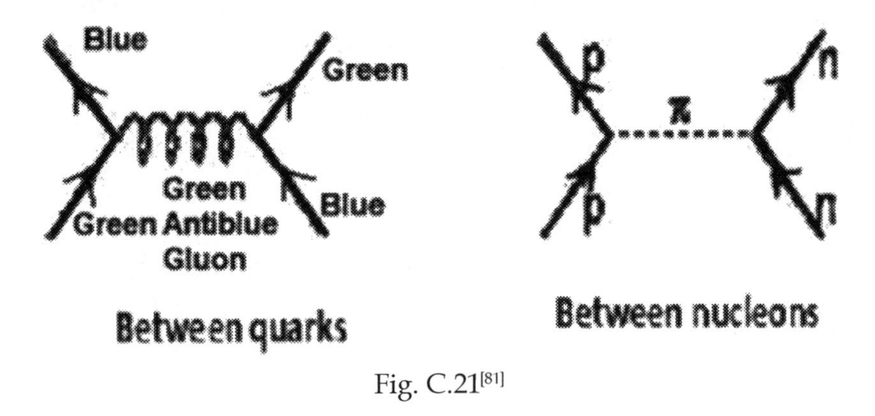

Fig. C.21[81]

APPENDIX-D

Copenhagen interpretation

Bohr first proposed Copenhagen interpretation in 1920. It says that a quantum particle doesn't exist in one state or another, but in all of its possible states at the same time. It is only when we observe its state, a quantum particle is forced to choose one probability, and that is its state we observe. Each time, a different observable state may be forced into and thus, the erratic behaviour of the quantum, can be explained by this interpretation.

The Copenhagen Interpretation has three basic postulations:

The first one—The wave function is a complete description of a wave or particle. The information which cannot be derived from the wave function does not exist. For example, a wave is covering a broad region, therefore does not possess a specific location.

The second one—The wave function of the wave/particle collapses when a measurement is done on it. For example: A wave packet is made of many waves each with its own momentum value. When the measurement is done on the wave packet, it reduces to a single wave and a single momentum.

The third one—If two properties of a particle like momentum and position have some uncertainty relation, no measurement can determine both properties simultaneously to a precision greater than the uncertainty relation allows. So, if we measure a wave/ particles location, its momentum becomes uncertain.[95][96]

The object's state of existing in all possible states at once is called its coherent superposition. Suppose the states of photon to be either as particle or wave at once forms the wave function of the object. When we observe the photon, the superposition collapses and the photon is forced into one of the states of its wave function.

In case of the example of Schrodinger's cat, of course, we know that this is not possible and nothing can be alive and dead at the same time. When we open the box, we "collapse the wave function" or "collapse the state" and have either a live cat or a dead cat. Let us now imagine a friend of ours is waiting outside when we open the box. For us, the wave function collapses and we see a live cat or a dead cat. For argument sake let us consider that we see the live cat. But our friend's wave function does not collapse until he comes into the room. So the friend waiting outside is still having perception of 50% possibility of cat being live or dead and he can then say that we owe our objective existence to this kind of observation by coming into the room and collapsing our state. This is explained by Heisenberg saying that "The wave function represents partly a fact and partly our knowledge of a fact." Thus Schrodinger's Cat reveals that our consciousness and knowledge are somehow mixed up in this process.

EPR Paradox and Bell's Inequality

For two entangled electrons, the state of one electron is determined by the other. The spin of an electron can either be up, say at state A, or down, at state B. In case of two entangled particles, if one is A, then the other must be B, and vice versa. But the state of one electron will remain unknown until the other one of them is observed,. Let us now consider that two such entangled electrons are generated through some process that gives opposite velocities to the pair. If one electron is observed and is found to be in state A, the state of the other electron will be found in state B at the very instant such observation takes place, irrespective of the distance between the two electrons. This implies that information can be instantaneously transmitted from one electron to its twin—faster than the speed of light![105]

Though he virtually invented the quantum theory of light, Quantum Mechanics did never attract Einstein's liking. The idea of quantum mechanics appeared to him as not workable and more or less he rejected the idea.

In fact while referring the uncertainty principle about position and momentum of the particle he quoted, "God does not play at dice with the universe." At that time Neils Bohr was also working with quantum mechanics and made a humorous comment, "Quit telling God what to do !"[100]

Einstein got two other like-minded physicists, Podolsky and Rosen and in 1935 wrote a famous paper entitled "Can Quantum-Mechanical Description of Physical Reality be Considered Complete?" They were of the opinion that the determination of the state of the electrons would be fixed at the moment of their creation. This suggestion so given by them was named the EPR Paradox, and it stood in direct opposition to quantum mechanics[105]. Thereafter, only in 1964, the proof in its favour was published by Bell in form of mathematical theorem which nicely proved that if momentum and position were absolute values (that is, they exists whether they were measured or not) then an inequality, would be satisfied which was later on called as Bell's Inequality. Einstein's position was firm as he quoted : "I think that a particle must have a separate reality independent of the measurements. That is an electron has spin, location and so forth even when it is not being measured. I like to think that the moon is there even if I am not looking at it ".[98][99]

But in the EPR paradox, Einstein and others imagined a scenario that would let them measure, for example, both the position and momentum of a particle with absolute certainty and that was a big no-no in quantum mechanics.

A perfect example is the case of the neutral pion. The pion is a subatomic particle having no spin which decays into two photons that shoot away from each other in opposite directions and the two photons are having spins opposite with each other. Now, Photons have spin, but these two photons came from a pion with no spin. So, if we know the spin of one photon, we can find out the spin of the other photon because their spins have to add up to no spin at all. As the photons came from a single pion, it is said that they are entangled. It is like an imaginary case where one basketball not spinning at all, suddenly degenerated into two golf balls, each spinning in opposite directions! [100]

Bell's Inequality

Bell's inequality says that

Number(A, not B) + Number(B, not C) >= Number(A, not C)

Example: Suppose in a class No. of boy students is A and no. of boys and girls students who have obtained above 80% marks is B. C is the no. of students having fair complexion.

Then the inequality means that the no. of boy students who have not got above 80% marks plus the no. of boys and girls who have got above 80% marks but not of fair complexion is greater than or equal to no. of boys who are not of fair complexion. This will be true for any value of A, B and C of the students. Now the question remains whether the state A and B of the entangled electron can be known in advance as contended by EPR?

Bell, proposed a method to investigate this question. Using photons, experiments were performed successfully and the results displayed a victory for quantum mechanics! The amazing relationship of entanglement really does occur and EPR was wrong. The long drawn debate was terminated. But still it 's implication is not fully understood whether the information can be passed faster than the speed of light.

ACKNOWLEDGEMENTS

1. The theory of everything by Hawking

2. Hyperspace by Kaku

3. Beyond Einstein by Kaku and Thompson

4. The large, the small and the human mind by Penrose

5. Theories of everything by J.D Barrow

6. The brief history of time-Hawking

7. Fundamental Particles, Fundamental Questions by Elizabeth H. Simmons

8. The tao of physics by Fritjof Capra

9. The Brihadaranyaka upanisad by Swami Krishnananda

10. Outlines of Indian philosophy by M. Hiriyanna

<u>**Internet resources**</u>

11. Big Bang: How Did the Universe Begin? by Yuki D. Takahashi

12. The elegant universe—Superstrings, Hidden Dimensions and the Quest for the Ultimate Theory—Brian Greene

13. Beyond the Big Bang-Inflation and the Very Early Universe—Gary Felder

14. Superstrings-John M. Pierre

15. A Universe of at Least 10 Dimensions-String Theory Finally Reconciles Theories of Relativity and Gravity-VIRGIL RENZULLI

16. Inflation for Beginners-John Gribbin

17. Einstein's Space time-James Overduin

18. The living universe—Duane Elgin

19. The dark matter mystery-Jupiter scientific

20. Particle Physics Timeline-Mountain Empire High School, California

21. Supersymmetry Phenomenology-Hitoshi Murayama

22. The particle adventure-the fundamental of matter and force-Particle data group, Lawrence Berkeley National Laboratory

23. Duality, Spacetime and Quantum Mechanics-Edward Witten

24. The Vedic Quest Concerning the Universe, Space and Time—Rati Saxena

25. Indian philosophy simplified-Multi AMSoft

26. Philosophical and Scientific Expositions of Creation of Universe Dr. Ashok Kumar Das & Dr. Saradamani Das

27. Aiterya Upanisad-Chanakya.com

28. Introduction to Isha Upanishad—Dr. C.S. Shah

29. Shabda Brahma-Dhrishti

30. Creation according to samkhya philosophy-Cosmic Lila

31. What Buddhists Believe—Sri Dhammananda Maha Thera

32. Acintita Sutta: Unconjecturable-Thanissaro Bhikkhu

33. Creation of the Universe by M.I. Liaqath Ali

34. The Origin of Life in the Universe: Buddhist Perspective-R M Jayatunge

35. Discussion on the origin of the universe—Raidou Hirota

36. Singularity—Realm of Gods By Gerald Shingleton

37. Islamic concept of creation of Universe Big bang and science-religion interaction by Prof. Faheem Ashraf

38. The Universe According to the Quran—Mumin Salih

39. Chhandogya Upanisads, Shankara Bhasya-India Divine.org

40. Universe as per Jainism-Parthiv Patel

41. The relative theory of cause and effect—Acharya Shree Mahapragyaji

42. Jainism And Non Creationism-Quaest.io

43. The Concept of God in Jainism—Mr. Devendra Kumar Jain

44. Concept of God in Jainism-Pravin K. Shah

45. Taoism and Confucianism-ushistory.org

46. Big Bang—The Bible Taught It First!—Dr. Hugh Ross

47. Is Creation Ex Nihilo A Post-Biblical Invention? An Examination Of Gerhard May's Proposal-Paul Copan

48. The Upper Paleolithic Revolution-New Archaeology

49. The Emergence and Evolution of Belief, Religion, and the Concept of God—F Quadri

50. Buddhism and the God-idea by Nyanaponika Thera

51. God in Buddhism-Buddhist tourism.com

52. Allah(God)—The Institute of Islamic Information and Education (III&E)

53. The pre-Islamic origin of "Allah"—Brother Andrew

54. Concept of God in Islam—WAMY Series

55. The god of Islam-answering-islam.org

56. Understanding the Trinity-Dick Tripp

57. The Christian Conception of God—R.A. Torrey

58. The Concept of God in the Vedas—Swami Tattwamayananda

59. Kasmai Devaya Havisha Vidhema (Rigveda)-Ashesh Joshi

60. Divine Message of the Veds—Pandit Shriram Sharma Acharya

61. Isopanisad-The Bhaktivedanta Book Trust International, Inc.

62. Ancient Physics History-Martyn Shuttleworth

63. Early Atomic Understanding-Main Time line

64. Ideas of physics—Dr. David Taylor

65. The Development of Quantum Mechanics—David M. Harrison.

66. Particle Physics Fundamentals, Introduction to Elementary Particle Physics-Andrew Zimmerman Jones

67. A Brief Introduction to Particle Physics—Department of Physics and Astronomy, University of Rochester

68. History of the atom from Democritus to Bohr and Schrodinger—Prof Mokeur

69. Modern atomic theory-N.De.Leon

70. Particle physics-Encyclopedia of science

71. The discovery of the Positron-Michael Miller

72. Atomic orbitals-Jim Clark

73. The standard Model-Ahren Sadoff

74. Pauli Exclusion Principle—Dan Sewell Ward

75. The evolution of Pauli's exclusion principle-Gordon N Fleming

76. Hideki Yukawa and the Pi Mesons—Thayer Watkins

77. The Standard Model-The Physics Hyper textbook

78. Standard model of particle physics-M.Woods

79. The official string theory website-Scientific American

80. The standard model-the universe(in the eyes of particle physics)-Particle physics, UK

81. Hyper physics-Georgia State University

82. Feynman's diagram-People's physics book

83. A guide to the nuclear wallchart

84. Say it with science-aps.org

85. CP violation-Excerpts from Encyclopedia Britanica

86. The Higgs Boson-Derek Robins

87. Higgs Boson—Encyclopaedia Britanica

88. The particle adventure-Particle data group of LB National laboratory

89. Subatomic particles-Encyclopaedia Britanica

90. Quantum Enigma by Bruce Rosenblum and Fred Kuttner

91. Consciousness and Quantum Mechanics Thomas J. McFarlane

92. Vedic quantum mechanics-Gurudev

93. Solving the quantum mysteries-John Gribbin

94. Schrodinger's cat—David.M.Harrison

95. Copenhagen Interpretation of Quantum Mechanics—Metaphysics Research Lab, CSLI, Stanford University

96. 21st century science-abyss.uoregon.edu

97. The Feynman Double Slit—John Blanton

98. Does Bell's Inequality rule out local theories of quantum mechanics?—John Blanton

99. The Einstein-Podolsk-Rosen Argument in Quantum Theory-Stanford Encyclopaedia of philosophy

100. Bell's Inequality and The EPR Paradox-Think Quest Team C008537

101. The myth of quantum consciousness by Victor. J.Stenger

102. Tat Tvam Asi-by Swami Sivananda

103. Erwin Schrödinger and the Upanishads by Michel Bitbol

104. The Quantum Brahman-A Study in the Relationship Between Quantum Theory and Vedic Cosmology—Robert E. Wilkinson

105. The Meaning of the Singularity: 1. A Single Particle Universe by Alon Retter

106. Modern Physics and Hindu Philosophy by Kashyap Vasavada

107. Quantum Physics and Eastern Philosophy by Tarja Kallio-Tamminen

108. Brahma Sutras-Swami Sivananda

109. Fundamentals of Special and General Relativity—K.D.Krori